VALUE-FREE
SCIENCE?

VALUE-FREE SCIENCE?

IDEALS AND ILLUSIONS

Edited by

Harold Kincaid
John Dupré
Alison Wylie

OXFORD
UNIVERSITY PRESS
2007

OXFORD
UNIVERSITY PRESS

Oxford University Press, Inc., publishes works that further
Oxford University's objective of excellence
in research, scholarship, and education.

Oxford   New York
Auckland   Cape Town   Dar es Salaam   Hong Kong   Karachi
Kuala Lumpur   Madrid   Melbourne   Mexico City   Nairobi
New Delhi   Shanghai   Taipei   Toronto

With offices in
Argentina   Austria   Brazil   Chile   Czech Republic   France   Greece
Guatemala   Hungary   Italy   Japan   Poland   Portugal   Singapore
South Korea   Switzerland   Thailand   Turkey   Ukraine   Vietnam

Published by Oxford University Press, Inc.
198 Madison Avenue, New York, New York 10016

www.oup.com

Oxford is a registered trademark of Oxford University Press

Library of Congress Cataloging-in-Publication Data
Value-free science? : ideals and illusions / edited by Harold Kincaid, John Dupré,
Alison Wylie.
   p. cm.
Includes index.
ISBN 978-0-19-530896-9
1. Science—Social aspects. 2. Science—Philosophy. 3. Objectivity.
I. Kincaid, Harold, 1952–   II. Dupré, John.   III. Wylie, Alison.
Q175.5.V35 2007
501—dc22   2006045302

Printed in the United States of America
on acid-free paper

Dedicated to the memory of Wesley Salmon

PREFACE

IT IS HARD TO HAVE BEEN INVOLVED IN DEBATES IN THE PHILOSOPHY OF
science in the last decades without confronting the place of values in
science. The three coeditors of this volume certainly have found this to
be the case in their own work. In Kincaid's treatment of confirmation
theory and issues in philosophy of the social sciences, in Dupré's investigation of questions about natural kinds in the biological and social sciences, and in Wylie's work on inference in archeology and on feminist
epistemology, questions of fact and value have proven to be close to the
surface. That conclusion is certainly reinforced for many other issues in
the philosophy of science by a broad range of others' research in science
studies. We are thus convinced that questions about the relation of fact
and value are crucial to debates that we care about. We also believe that
(1) issues of fact and value have too often been left implicit rather than
confronted head on and (2) that there are defensible positions other
than the dichotomous views that good science must always be value free
and that all science is politics.

These experiences and convictions led Kincaid to organize in 2002 a
conference with same title as this volume at the University of Alabama at
Birmingham and led Dupré and Wylie to contribute papers to that conference. The conference included both invited and submitted papers

and drew a large and enthusiatic set of presenters. Most of the papers in this volume originated from those presentations.

This book is dedicated to the memory of Wes Salmon, who was a friend to several of us and much respected as a philosopher of science by all of us. We think this is fitting for a book on values and science, not because Wes would have agreed with all or even much that is written here but because we know that he took these issues seriously. At the conference that initiated this volume, Wes was an invited speaker. Unfortunately, he died shortly thereafter, without having had time to write up his contribution. His talk built on his paper "Thomas Kuhn Meets Thomas Bayes," with the theme being the complexity of considerations in confirmation. Unfortunately, we were not prescient enough to take careful notes.

However, we do have some written evidence of his final conclusions because of the following story about Wes, which says as much about his character as about his views on values and science. The story is this. The conference in which Wes participated drew enough interest to result in simultaneous sessions, something the organizers had not anticipated. As a result, logistics for meeting spaces were complicated. Those logistics were being overseen by Fletcher Harvey, a longtime audiovisual coordinator for programs in the biomedical sciences and, not surprisingly, unfamiliar with who is who in the philosophy of science. At one particularly chaotic moment between sessions, Fletcher grabbed Wes by the shoulders and ordered him to block entry into a conference room until he said otherwise. Being the gentleman he was, Wes complied without complaint. Later, Fletcher learned Wes's identity and in admiration (and probably embarrassment) had Wes autograph the program for the conference. That signed copy is now framed and hangs in the main conference room Harvey manages—along with other such signed programs, several by Nobel laureates. From that program, we know something of Wes's views on values and science: It is signed "Value Free Science=Worthless Science. Wes Salmon."

We would like to acknowledge the National Science Foundation for grant number NSF 0080217, as well as the Center for Ethics and Values in the Sciences and the Department of Philosophy at the University of Alabama at Birmingham for financial support of the conference that started this volume, and the participants of the conference for their many helpful discussions and ongoing moral support. Two anonymous referees also made useful comments on content and organization, and we would like to thank them for their efforts. Dupré would like to acknowledge the support of the Economic and Social Research Council. His contribution to this volume was part of the work of the ESRC Centre for Genomics in Society.

CONTENTS

CONTRIBUTORS

Gerald Doppelt is Professor of Philosophy and Director of the Graduate Program in History, Philosophy, and Sociology of Science at the University of California at San Diego.

Heather Douglas is Assistant Professor of Philosophy at the University of Tennessee.

John Dupré is Professor of Philosophy of Science and Director of the ESRC Centre for Genomics in Society at Exeter.

Harold Kincaid is Professor and Chair of the Department of Philosophy and Director of the Center for Ethics and Values in the Sciences at the University of Alabama at Birmingham.

Lynn Hankinson Nelson is Professor of Philosophy at the University of Washington.

John T. Roberts is Associate Professor of Philosophy at the University of North Carolina at Chapel Hill.

Michael Root is Professor of Philosophy at the University of Minnesota.

Sherrilyn Roush is Associate Professor of Philosophy at the University of California at Berkeley.

Elliott Sober is Hans Reichenbach Professor and William F. Vilas Research Professor in the Department of Philosophy at the University of Wisconsin.

K. Brad Wray is Associate Professor of Philosophy at the State University of New York at Oswego.

Alison Wylie is Professor of Philosophy at the University of Washington.

VALUE-FREE
SCIENCE?

INTRODUCTION

ALL THE CHAPTERS IN THIS BOOK RAISE DOUBTS ABOUT THE IDEAL OF A value-free science. That ideal takes science to be objective and rational and to tell us about the way things are, but not the way they should be. That ideal has dominated our conception of science for centuries. Critics of the value-free ideal are clearly challenging deeply held assumptions about a key institution of the modern era. This introduction explains in the first two sections why we should care about such questions and outlines the history of these questions, leaving the notion of value freedom at the intuitive level. The next section then seeks to sort out the numerous different theses involved in the idea of a value-free science and to sketch some general kinds of arguments offered for and against those claims. The last section then relates the chapters of this book to these issues and arguments.

Why Care?

The place of values in science is an important question that cuts across numerous debates in the philosophy, history, and social studies of science. Debates over the nature of scientific rationality and scientific change, scientific realism, the prospects for a theory of confirmation, the role of gender in science, and much more are tied up with questions about the place of values in science. However, although whether science can and should be value free is an interesting intellectual question, its interest is not just intellectual. Science has a fundamental place in Western societies. The practice of medicine depends on on it. Governmental policy—on the safety of drugs, the effects of inflation, or the threat posed by global warming—is influenced by science. And fundamental ideals of Western society such as rationality and progress are grounded in certain conceptions of science. So when the value freedom of science is questioned, a fundamental institution in our lives is being challenged. Seeing those challenges explains why we should care about the claim that science is value free.

The value-free ideal sees science as neutral: Scientific claims are true or false and depend on evidence independent of anyone's moral or political views, and though the correct application of moral or political values may depend on what the facts are, the appropriateness of such values is a matter quite independent of facts about the world. The job of science is to tell us the facts. Moral and political decisions, on the other hand, are matters of values and preferences. The two domains are autonomous.

If the critics of the value-free science ideal are right, then these traditional claims about science not only are ungrounded but also can have pernicious consequences. If the *content* of science—not just its application[1]—can and must involve values, then presenting scientific results as entirely neutral is deceptive. It means ignoring the value assumptions that go into science and the value implications of scientific results. Important value assumptions will be hidden behind a cloak of neutrality in public debates over policy and morality. If scientific results concerning IQ and race, free markets and growth, or environmental emissions and planetary weather make value assumptions, treating them as entirely neutral is misleading at best.

Doubts about value-free science call for similar changes in how we conceive the obligations of scientists and the public. If science makes value commitments, then scientists are responsible for those commitments—for making them explicit and for considering their consequences. Moreover, it is not simply the obligations of scientists that are at issue. If values play a role in science, then the public and public officials cannot take scientific results as given and scientific authorities as beyond challenge. Responsible public policy will require responsible use of science; responsible use of science will require explicit critical awareness of its value assumptions.

Finally, if we do reject the ideal of value-free science, what will replace it? Advocates of value freedom believe they are defending objective science free from bias and corruption. How are these ideals to be realized if science is value laden? If they are not to be abandoned, how might the practice and assessment of science be affected?

We should care not only about the role of values in science but also about the role of science in values. Treating values as entirely outside the realm of scientific rationality and objectivity can easily promote the idea that moral matters are entirely subjective and arational. Those views, in turn, have tremendous implications for how we should conduct moral debate and think of moral obligation. If values are an essential part of good science, and the content and justification for values, in turn, are sometimes sensitive to the deliverances of science, then moral issues must be differently conceived. The place of values in science thus has practical implications as deep as its intellectual implications.

The History

The ideal of a value-free science is intimately tied up with a corresponding attitude to morality and values—the attitude that values are subjective and unamenable to rational argumentation. The current debate over values in science has its origins in reaction to that idea. This section sketches those historical origins and the responses to them.

Questions about the relation between facts and values have been raised since the beginning of Western philosophy. But for the modern debate, it is Hume who defined the issues (1888, 469) with his claim that *ought* cannot be derived from *is* or, in other words, values cannot be deduced from facts. By *deduced*, Hume meant logical deduction: Statements of fact on their own never logically entail any moral conclusion. If the premises of an argument make no moral claims, then the conclusion cannot either. Every argument that seems to derive a moral conclusion must have a disguised moral premise.

This sets a sharp divide between facts and values. We cannot deduce moral results by reasoning from empirical observations of objective events. But deducing conclusions from facts is the epitome of rationality and a hallmark of science. Morality, then, is a different sort of enterprise, one that is essentially about human sentiments. Moral or value judgments are neither true nor false. For Hume, all truths are one of two kinds—truths of reason or truths of fact. Mathematical or logical truths fall into the first category, ordinary empirical assertions into the latter. Moral claims fall into neither.

Hume's divide between *ought* and *is*, fact and value, reverberates throughout all of Western intellectual history, even when the rationality of morality is defended. In particular, Kant rejected Hume's claim that morality involves the expression of sentiment and argued that basic moral principles are principles of rationality. Yet in defending morality, he agreed that moral claims are neither truths of reason nor truths of fact. Instead, they are synthetic a priori truths—truths that can be known independent of experience and yet are not mere logical truths. Like the claim that "every effect has a cause," moral truths are of very special sort that are not warranted by ordinary experience but imposed by the human mind. Values are thus still quite different from ordinary facts.

After Kant, the decisive period for the fact-value distinction, particularly as it relates to science, was the logical positivism movement. The positivists agreed with Kant that moral and metaphysical claims were neither truths of reason nor truths of fact. However, they denied that there was any such thing as synthetic a priori truth and thus held with Hume that moral judgments are neither true nor false, unlike judgments about facts.

The classic logical positivist statement of this view was that of Ayer (1952). Moral or value claims do not state propositions that can be true or false. The content of moral claims is instead emotive; moral and value claims express approval or disapproval. To say "X is immoral" is to say something like "Boo hiss, X."

Variants of Ayer's view predominated up through the 1950s. Stevenson (1960) thought that the meaning of moral claims was captured by approval and recommendation; "X is moral or valuable" is equivalent to saying "I approve of X; do so as well." Moral and value claims were once again neither true nor false and thus fundamentally different than factual statements.

Worries about the fact-value distinction and its implications came from multiple sources in the 1950s and led fairly directly to the concerns discussed in this volume.[2] Doubts come from at least four sources: difficulties in explaining the place of argument in ethics on the emotivist view, doubts about the analytic-synthetic or truths of reasons–truths of facts distinction stemming from Quine's work, apparent cases where language has both evaluative and factual components, and apparent cases in science where values seem to have a fundamental role.

Although there may be differences between moral discourse and other language, there are also deep similarities. It is hard to see how to capture those similarities if moral claims are expressions of emotion, sentiment, or approval. The similarity is that moral claims are used in arguments apparently as are factual claims. People give reasons for their moral judgments, and those reasons can take the form of formal arguments. So opponents of capital punishment can ground their judgment in the argument: "If capital punishment is applied unequally, it is unfair. But capital punishment is applied unequally by race, sex, and economic status. Therefore, capital punishment is unfair." It seems as if this conclusion follows deductively from these premises. Yet if moral claims are neither true nor false, the standard notion of deductively valid arguments—that the truth of the premises guarantees the truth of the conclusion—should have no application. Similarly, the claim that "if capital punishment is applied unequally, it is unfair" does not say "Boo, hiss capital punishment" but seems to assert a relation between two propositions. How this can be rendered as an expression of emotion is not at all clear. Finally, to put the point in less technical and more intuitive terms, we give reasons for our moral beliefs in a way that would make no sense for mere expressions of emotion. Facts and values don't look so different.

Ordinary language philosophers in the 1950s and 1960s also noted that a great deal of ordinary language was not easily divided into the straightforwardly factual or evaluative. Examples can be drawn from quite mundane areas of everyday practice, such as the grading of commodities. To describe an apple as "extra fancy" is pretty clearly to provide an evaluation of it. But if one consults the relevant criteria laid down by the appropriate authorities, they are factually quite precise, specifying exact requirements of size, color, and so on. It is also helpful to move away from the broad general terms characteristic of so much philosophy and consider the much more specific terms in which most everyday evaluation takes place. As J. L. Austin memorably suggested, it is helpful to turn one's attention from the beautiful to the dainty and the dumpy. It is, on the whole, a good thing to be dainty, but the factual criteria for this particular aesthetic achievement are fairly precise. Similarly, it is an aesthetic flaw to be dumpy, but it is a risk the slender avoid, whatever their other aesthetic strengths and weaknesses.

A third development that indirectly led to many doubts about the distinction was due to Quine, in particular, his holism. Quine argued that our beliefs constitute a web where every belief is at least indirectly tied to all others by logical and other evidential connections. This means, on Quine's view, that there is no sharp way to separate our beliefs into those that are truths of reason and those that are truths of fact. In science and elsewhere, belief revision is not dictated by algorithmic rules of reason but is a piecemeal process guided in part by pragmatic factors.

These Quinean ideas have a number of implications about values and facts. First is that it deepens the significance of the hybrid cases just discussed. Terms such as *dainty* and *dumpy* have semantic links to both evaluative claims (it's better not to look dumpy) and factual claims (to be dumpy, one must be on the plump side). If the meaning of terms is at least in part determined by the various connections a sentence has to other sentences in the language, then we should expect meaning to be holistic—to depend on a term's use in many different kinds of sentences. So it is quite unsurprising that a given term might have connections to both our factual and value judgments, precisely what Austin is pointing to. Moreover, the phenomena Austin identifies should not be rare, and it is easy enough to find further examples. Analyses of racial and gender categories in the social and biomedical sciences, for example, clearly show both evaluative and factual components. Hilary Putnam (2002) has explicitly used these Quinean arguments to show that fact and value are "entangled."

Quine's doctrines also motivated much more careful looks at the actual practice of science. If revising the scientific web of belief is not a mechanical and purely logical process, then how actual scientists reason and pursue research becomes relevant for our understanding of how science works. Extralogical considerations must mediate scientific inference, thereby opening a wedge to value and pragmatic considerations. This change in focus has many ramifications for how we see fact and value in science.

Some of the earliest examples of this change in focus are found in arguments from Rudner (1953) and Myrdal (1970) that still have force today. In the actual practice of science, it is often not the case that experimental results definitely confirm or disconfirm a hypothesis. Instead, evidence piles up gradually in favor of some particular claim. Rudner pointed out that therefore the decision to accept a scientific result requires a sense of both how much evidence is at hand and how much certainty we need. If there are large costs from being wrong, then just having some evidence may not be enough. Deciding the relevant costs, however, invokes human values. So deciding when to accept a scientific hypothesis essentially involves values.

Myrdal argued that a second aspect of science where values surface concerns deciding what to study. This is not just a matter of human interests steering some people to biology and others to history, though it is that, too. Myrdal was interested in the economics of development. He noted that the economics of his time—and the situation has not changed entirely even today—asked questions about what would happen in equilibrium, that is, the situation where all sectors of the economy produce and demand just as much as is demanded and produced by all others. This emphasis fit with the political philosophy of most economists, who saw a just distribution of income as one resulting from choices in a market where no one could be made better off without making someone else worse off. Markets in equilibrium have these traits (given certain other assumptions), and economists tend to refer most economic problems to the equilibrium standard. Yet for Myrdal, the economic underdevelopment in large parts of the world was not an equilibrium phenomenon, and thus the focus on equilibrium meant not asking important questions. Values were influencing the content of social science.

By far the biggest Quinean impact came in 1962, with Thomas Kuhn's *Structure of Scientific Revolutions*. Kuhn was pursuing the task of how science actually works. He was also much influenced by Quine's holism and emphasis on pragmatic factors in theory choice. Science, on Kuhn's view, is guided by paradigms, where theory, methods, and data are interwoven in a tight web, one so tight that changes in paradigms are radical changes. That change comes about when problems cannot be solved and a new paradigm promising solutions gains sufficient adherents to dominate the journals, universities, and other scientific institutions.

Scientific change is arational, and sociological processes are part and parcel of science.

After Kuhn, a flood of studies in history, philosophy, and sociology of science showed how interests, values, cultural ideals, and the like influence even the hardest of sciences. The critics of emotivism who defended the rationality of values had been joined from the opposite end by a critic who identified the values in rationality.

The Issues

The claim that values influence science is much like the claim that genes influence behavior: At this level of vagueness, almost no one will disagree, but there are underlying deep disagreements about how, how much, and with what ramifications they do so, once more specific theses are asserted. This section distinguishes various ways that science might be value laden and their implications. The claim that science is value laden can be separated into four different dimensions: the kinds of values involved, how they are involved, where they are involved, and what effect their involvement has. With those distinctions in hand, we can make an initial pass at the kinds of arguments given and what they may or may not show.

"Values" in these debates generally are those things that individuals think are worthy of being or ought to be promoted, advanced, and realized. In this sense, then, "values" refers not just to moral, ethical, or political values, which themselves may be distinct. In particular, a distinction is often made between epistemic and nonepistemic values. Epistemic values are such things as truth or knowledge. Truth might be valued because of its moral consequences, but it can be valued for its own sake and thus seems to be a further kind of value beyond ethical or political ones.

These distinctions are important because they apparently ground rather different theses. Science might involve political values—say, in the funding process—but not moral values, and vice versa. More important, science might involve epistemic values without thereby invoking moral or political ones. So we might hope that scientists value truth and knowledge and grant that science in that sense is value laden. We might likewise note that some scientists put a premium on predictive accuracy and others on explanatory scope and in that way grant that science is value laden. But that is seemingly a very different thesis from the claim that moral or political views are involved in science. We say "seemingly" because whether there is a viable epistemic-nonepistemic distinction is one of the issues under debate.

Putting these distinctions aside, the second question to ask is how values are involved. Multiple dimensions are relevant here. Is the involvement inevitable or only possible? Finding that moral values are inevitably involved in science is a considerably more consequential finding than that they sometimes can be. The idea that values are only sometimes involved leads easily to the thought that this is a circumstance that we might take steps to prevent. And this, in turn, leads to the idea that science in which values are involved is, to that extent, bad science. The notion that only bad science involves values is a weak and perhaps uncontentious version of the value ladenness of science.

Another relevant dichotomy is between values being involved by implication or in the use of science as opposed to by presupposition or in the construction of science. Finding that human fetuses consciously experience pain only at approximately six months after gestation may have serious moral implications about abortion if you believe that the ability to feel pain is crucial to having interests and rights; the finding itself presumably could be done without making any assumptions about those moral issues. So values are implied by such a fact but not presupposed by it. To put this another way, in the latter case, values are part of the content of the science; in the former case, they are not. In terms of our previous distinction between essential and nonessential use of values, values could be essential in that they were always present in some form (e.g., in scientists' motivations) but still not part of the content of science.

The difference here is important. When moral values are presupposed by science, there may be a worry about the objectivity of the results, at least to the extent that there are worries about the objectivity of morality and values. A science that has moral *implications* need not raise such doubts, if the implications come from moral values clearly outside the science itself.

A third question to ask is where in the scientific arena values are involved. Science is a complex enterprise, and we should not expect that finding values should have the same implications regardless of where they are found. For analytic purposes, let us distinguish three broad parts of the scientific enterprise: the areas investigated, the hypotheses considered and the evidence that is taken to show one of those hypotheses to be confirmed or superior, and the use of scientific results to produce explanations. (These aspects may be interconnected, and how and why they are is a further set of issues in the debate over values in science.)

Consider first the areas investigated. The prospect of value-laden science shows up in several ways here. Many people have argued that some domains can be investigated in a value-free way and that others in principle cannot, with the social sciences usually being cited as in the second category. Values can also obviously influence science by dictating what is studied—the questions asked. There are, for example, explicit procedures in government funding agencies to determine priorities, and they strongly influence what gets studied. More fundamentally, values might be involved in deciding what categories or basic objects constitute the area to be studied; picking out an area to investigate and deciding what basic entities and categories characterize it are mutually interacting processes. To use an example from John Dupré in chapter 1, deciding that there is "rape" among ducks is not simply read off the facts and may well involve values in some way.

Values can also be appealed to in constructing scientific explanations. If we take the citing of causes as a paradigm case of explanation, then this is easy to see. We frequently identify some one thing as "the" cause of a phenomenon. But every effect is the result of a complex of factors. Which factor we focus on may not be dictated by the facts but by our interests. So, for example, in explaining crime, defenders of libertarian accounts of justice are likely to focus on individual-level causes of crime. Those with more social democratic leanings are likely to emphasize the social causes of crime. Values are determining how the explanation goes. These kinds of cases have been examined in the work of Longino (1990, 2002) in some detail.

Values in the confirmation process and in the selection of hypotheses to consider would put values at the heart of science. If what we think ought to be the case is an essential part of the evidence—if it is presupposed rather than just implicated, to use our earlier terminology—then we have value-laden science in one of the most thoroughgoing senses. If values dictate which hypotheses are live options and seriously considered, they obviously influence the outcome of inquiry in a much stronger way than they would, say, if they influenced the areas of science pursued.

Finally, we have to ask what follows if science is value laden. Does it show that science is biased? Not factual? Subjective? Or is the presence of values compatible with good science—science that is objective, well confirmed, and so forth? It should be obvious by now that this question is unlikely to have one answer, because values can be involved in the many different ways just described. We should not expect that involvement to have the same implications for assessing science in what are very different contexts. Thus there are many possible positions at stake.

Given this set of dimensions, it will be useful to state the value-free ideal in its strongest forms. Consider the following three claims:

1. Objective, well-confirmed science never essentially presupposes nonepistemic values in determining what the evidence is or how strong it is.
2. Objective, well-confirmed science never essentially presupposes nonepistemic values in providing and assessing the epistemic status of explanations.
3. Objective, well-confirmed science never essentially presupposes nonepistemic values in determining the problems scientists address.

All three claims deny that nonepistemic values are logically presupposed by the content of science. This is compatible with many other weaker forms of value ladenness; for example, epistemic values may be presupposed or nonepistemic values may be consistently present in various ways, but they are not essential to the content of the science. The staunchest defender of value freedom would assert all three claims. Many have defended value freedom in the sense of claim 1 but granted value ladenness in the sense of claims 2 and 3 on the grounds that choices of explanatory factors and intellectual topics have an essential pragmatic element that providing evidence does not.

Does anyone actually hold these strong views about value freedom? We think they are an ideal implicit in much that is said about values and science, but they are also explicitly defended as well. A good case in point are the articles in *Scrutinizing Feminist Epistemology* (Pinnick, Koertge, and Almeder 2003). There we find the following thoughts:

> The argument seems to be that the idea that feminist values couldn't constitute evidence with respect to this or that theory rests on an untenable distinction of descriptive vs. normative. . . . What is at issue is not whether moral or political criticisms of priorities within science . . . are ever appropriate . . . but whether it is possible to derive an "is" from an "ought." . . . That it is false is manifest as soon as you express it plainly: that propositions about what states of affairs are desirable or deplorable could never be evidence that things are, or are not, so. (Haack 2003, p. 13)

> On this model, the experiments that scientists conduct may well be limited by both technological and ethical factors, but no restrictions are to be placed on the intellectual content of either the problems scientists address or the answers they explore. (Koertge 2003, p. 225)

Clearly Haack is denying that nonepistemic values can have any essential place in determining what the evidence is. Koertge's discussion is obscure (what it is "to place restrictions on the intellectual content of a problem" is left unexplicated), but it is not hard to see her asserting claim 2 and perhaps even claim 3.

Having argued that the question "Is science value free?" is really many different questions, we next want to briefly sketch some of the more common types of arguments answering those questions. Three types of arguments are commonly made for the value-laden nature of science: (1) arguments from denying the distinction between fact and value, (2) arguments from underdetermination, and (3) arguments from the social processes of science. These arguments take different forms in different authors, and we do not pretend to assess them in detail. But some awareness of these argument forms and their analysis will be helpful in reading the chapters that follow.

One natural way to argue against the value-free science ideal is to attack the distinction between facts and values—to argue that the claim is misguided from the beginning. There are various ways to support such an attack. Direct counterexamples are one strategy: Find cases of scientific investigation where the claims made are both evaluative and factual. The examples of hybrid terminology previously cited are an example. A more systematic strategy is to offer theoretical reasons that fact and value are not independent. Arguments from the holism of meaning— mentioned in the discussion of Quine—pursue this tack.

A second set of arguments points to underdetermination of two sorts: underdetermination of theory by data and underdetermination of theory choice by epistemic values. Underdetermination of theory by data is the idea that once we have all the data, there may be multiple hypotheses compatible with the data. Duhem (1991) and Quine (Quine and Ullian 1978) argued that this must be the case because of the holism of testing: We can always revise different parts of the web of belief in the face of new evidence. Intuitive examples of underdetermination come from "curve-fitting" problems: Given a set of data points, there are infinitely many ways to continue whatever trend you identify in the data.

Underdetermination of theory choice by epistemic values—such as scope or accuracy—was one of Kuhn's main claims. If two scientists with different theories agree on what epistemic virtues a theory ought to have and on the data, they may nonetheless not agree on which theory is best supported, because they may rank epistemic values differently. A theory that maximizes predictive accuracy might fare less well in terms of scope compared with another that did less well in terms of accuracy. Values can be traded off, resulting in a standoff.

Both kinds of underdetermination seem to support the value ladenness of science. The evidence or the evidence plus the epistemic values do not tell us what to believe. So values either can or must be involved; Longino, for example, seems at times to defend both (1990, 2002). Because these are supposed to be essential traits of science, we have an argument that nonepistemic values are essentially involved in good science.

A third type of argument sees value ladenness in the social processes of science. Here, a plethora of post-Kuhnian studies detail how individuals interact to produce scientific outcomes and how interests and values are involved in those processes. One simple but powerful example comes from the cognitive division of labor in science: Many particle physics experiments are so large and complex that no one individual can grasp and verify all the details, which makes trust an essential component. That raises the prospect, however, that in assessing the evidence, scientists have to make value judgments about character.

Four initial questions confront all three arguments for value ladenness: Are the values involved nonepistemic as opposed to epistemic? Do the arguments show that science must involve values or only that it can? Are values presupposed, or are evaluative conclusions only implied? Do the arguments show that good science can involve values or only that bad science—that is, biased science—can? An affirmative answer on all four questions will usually mean the argument was quite significant, if successful. Significance diminishes as the number of positive answers declines, for we approach less controversial senses of value ladenness as we do.

Beginning with the no distinction argument, negative answers might be raised by various routes. If there are cases of overlap—cases where something seems both factual and evaluative—we might still think there are clear cases where the two are distinct. There are gray cases between being bald and not, but we do not conclude there is no difference in general. If there are clear cases, then it may be that the distinction is valid and that some device must be found to handle the exceptions. So we might try to partition the meaning of "courage" into a factual component (e.g., overcoming fear in the face of danger) and a value component, such as behaving in a praiseworthy manner in the face of danger. Critics on the other side will, of course, then challenge whether such partitions work in specific cases or have other unwanted implications.

Another route to negative answers would be to undermine the theoretical arguments challenging the distinction between fact and value. Perhaps claims about the holism of meaning are just mistaken. Or perhaps the holism is less drastic than depicted, and we can partition meanings as suggested for the cases where holism is present.

The underdetermination arguments, of course, go nowhere if the alleged underdetermination is suspect. Questions have indeed been raised whether these Quinean and Kuhnian doctrines are compelling. Further issues concern exactly how values gain entry if underdetermination does obtain. Why do moral rather than scientific values break the ties? If two theories are tied on the evidence, why not pick the one that seems to offer the greatest promise of fruitful future research, for example? Maybe the best response to equally warranted competitors is to withhold judgment. After all, it is just such situations that motivate antirealist doubts about theoretical entities.

Social constructionist arguments will fail to show significant conclusions if the role of values in any specific instance is just an instance of biased, bad science. Then values will have no inevitable connection to science. Arguments to the effect that science is in some sense essentially social deny that values are contingently involved, because social processes inevitably bring interests and decisions about how collective decisions are to be made. Interesting questions then arise about objectivity: Is science objective despite the presence of values? Or are values necessary for the objectivity of science? The answers here get more interesting and controversial as the values involved move from epistemic to more obviously moral and political ones.

The Chapters and Their Arguments

The chapters in this book fall into three broad categories: case studies, discussions of the evidential relations between fact and value, and examination of the fact-value distinction from specific programs in the philosophy of science. Although nearly all of the chapters discuss specific examples, the papers by Dupré, Root, Wylie and Nelson, and Wray are detailed studies of specific scientific controversies and practices. Dupré (chapter 1) looks at the concept of rape in evolutionary psychology and inflation in economics, Root (chapter 2) at the sociological study of social problems, Wylie and Nelson (chapter 3) at the role of feminist values in archeology and developmental biology, and Wray (chapter 4) at values in the peer review process of science. Sober (chapter 5) and Douglas (chapter 6) fall into the second category: They examine the evidential interrelations between value claims and fact claims. Roberts, Roush, Doppelt, and Kincaid examine the issues from particular perspectives in the philosophy of science—Roberts (chapter 7) from the perspective of logical empiricism, Roush (chapter 8) from van Fraassen's constructive empiricism, Doppelt (chapter 9) from a post-Kuhnian standpoint that he has been developing over a number of years, and Kincaid (chapter 10) from a general approach in epistemology labeled contextualism.

Though there is not complete consensus, some major themes include:

> Facts and values can mutually support each other in science.
> The ideal of a value-free science can be bad for science.
> The role of epistemic values in science is considerably more complex and contested than is generally recognized.
> In the human sciences, values have an especially central role.

> Scientific reasoning and objectivity are not captured by purely mechanical rules and are compatible with and sometimes enhanced by the presence of values.

These are conclusions. The interest of the chapters is in the arguments given for them.

Dupré uses the concept of rape in evolutionary psychology and work and inflation in economics to argue that sometimes values are centrally and necessarily involved in doing relevant science. These concepts essentially involve both descriptive and normative elements. Any attempt to eliminate the normative elements will result in not explaining the phenomena we are interested in but something else. Dupré suggests that this is a general moral — that when science is directed toward our interests, it can and must be value laden.

Root argues for similar conclusions. He does so by tracing the controversy in sociology over the study of social problems running from Weber (1968) to Merton (1966) and Blumer (1971) up to the present. All three wanted to study social problems — such as spousal abuse — in a value-free way. For Merton, social problems result from the violation of norms, but the norms of society, not those of the investigator. However, those norms need not be consciously acknowledged by the individual involved and may be latent or implicit. Blumer objects that this is not sufficiently value-free social science, for it ignores the fact that social problems are socially constructed; particular conditions are labeled social problems by society itself, and invoking latent values ignores this and results in the investigator imposing values. More recently, Woolgar and Pawluch (1985) have gone even further to claim that even "conditions" are socially constructed. Root uses this dialectic to argue that social science explanations sometimes have to rest on judgments about what individuals would do or believe if they had all the relevant information. But to do so, social scientists have to rely on what they take to be reasonable moral claims. You have to have a sense of what is unjustifiable behavior to study spousal abuse. This is Dupré's conclusion, and like Dupré, Root believes that values are thus shown essential to good science, not that we can say nothing objective about spousal abuse.

Wylie and Nelson argue that feminist values have been central to producing better — more objective — science in several specific instances.

They point out that objective individuals—ones with no explicit moral stance—need not produce objective knowledge, and those with a committed stance may produce more objective results, a point made with some generality recently by Solomon (2002). They take this to show that objective, value-free science and science as politics are not the only possibilities, contra many commentators.

Wylie and Nelson support their claims with two detailed studies. The first shows that the introduction into archeology of greater numbers of women with commitments to feminist principles led to the production of empirically better science in a number of instances. The second case study comes from developmental biology. There, those with feminist standpoints argued for a differing trade-off of epistemic virtues such as generality, simplicity, and explanatory power than that employed in dominant models of development. This is an instance of the kind of complexity of epistemic norms and their interaction with values that Doppelt emphasizes in his chapter.

Finally, Wray surveys what is known from empirical studies about science on the role of nonepistemic values. In particular, he looks at their effects on the judgments scientists make about each other: explicit peer review ratings in funding decisions and implicit judgments in deciding whose work to cite. The evidence seems to show that nonepistemic values, such as gender bias, influence funding evaluations significantly but have no perceptible influence on citations. One possible explanation is that scientists need the work of others to promote their interests in research, and gender bias is overruled; the incentives in the funding case are entirely different with different results. Wray's chapter thus illustrates the point made earlier, that values can be involved in diverse aspects of science and need not have a uniform effect.

The chapters by Sober and Douglas in part II focus on values and scientific evidence. One way to put the ideal of value freedom, according to Sober, is "believing that a proposition has good or bad ethical consequences is not evidence for its truth." He is suspicious on the general grounds that we can't tell a priori what is evidence for what (a point fundamental to the arguments in Kincaid's chapter) and constructs some compelling counterexamples. The evidence relation is a three-place one between data, hypothesis, and background knowledge, thus allowing background knowledge to link fact and value. This makes possible evidential reasoning from *is* to *ought* and vice versa, an idea expressed in some form by many of the other chapters. However, Sober argues that there is a qualified sense in which the value-free ideal is right: Judgments about the moral consequences of a proposition cannot provide new information about its truth.

Douglas would probably disagree with even this more moderate thesis. She argues that the value-free ideal, even in the process of confirmation, can be bad for science, and seeing how can help us better understand scientific controversies. Her main focus is science used for public policy, and her argument is a generalization of Rudner's argument that we have to consider the value consequences of accepting scientific theories, in addition to their probability. However, she takes the argument a step further. Rudner seems to leave us with a fact-value distinction in the end—probabilities of truth are separate from consequences. Douglas argues that real science is much messier, for in coming to estimates of the probabilities themselves, scientists have to make choices that have moral consequences, so values are already there. She illustrates this claim with examples from risk analysis, where value consequences are present in categorizing rat liver cells as cancerous.

Douglas draws some interesting morals about scientists' ethical values in producing science for public policy. Scientists should make value judgments, and it would be a disaster if they tried to leave all such judgments to policy makers or the public for two related reasons. First, it simply is not possible to pass on the raw data denuded of value judgments, as the rat liver cancer case illustrates. Second, policy makers do not have the requisite knowledge to know what consequences of the science we should be worried about. And given that value consequences run all the way down through the science, policy makers would have to be involved in all aspects of scientific research. Douglas is right to think that this would breed a huge bureaucracy and hinder science. Scientists themselves have to take moral responsibility.

The final four chapters of part III ask how the value-free ideal looks from the perspective of different theoretical stands in philosophy of science. Roberts pursues the issues from the vantage point of logical empiricism. There are good reasons to do so. The aspect of logical empiricism he focuses on—the ideal of a formal confirmation theory—retains a powerful sway on the thinking of philosophers and scientists alike. Moreover, there has been a virtual renaissance of literature on the positivists that argues that their views were subtler and more diverse than the common slanderous use of the term *positivist* allows. And Roberts believes there are general morals in his analysis.

Although his argument has multiple steps, its basic gist is this: Philosophy of science for the logical empiricist is not about producing true statements concerning science but, instead, about producing useful conceptual tools—proposals for how to understand notions like "well confirmed." Those proposals cannot currently be justified by showing they promote our epistemic ends, for we don't have the requisite empirical knowledge. The only available justification thus has to shift from the consequences of the proposal to its inherent value (in the metaethics lingo, from consequentialist to deontological). The logical empiricists have such a justification, namely, that their proposal to focus on purely formal aspects and to ignore the content of theories instantiates an important value—egalitarianism of worldviews. Thus logical empiricism needs to reject the value-free ideal rather than advocate it.

In this argument, we find further evidence for the point made earlier that values can be involved in many ways. Roberts's distinction between science that promotes certain values versus science that instantiates them adds to our understanding and deserves further study.

Roush argues that constructive empiricism allows a place for values without necessarily committing us to the idea that values provide evidence about factual claims (a strong version of which was criticized by Sober). Constructive empiricism holds that the aim of scientific theory is empirical adequacy, that is, fit with observations. However, it is not unusual for competing theories to be "tied" when it comes to empirical adequacy. This is not the claim that theories must be underdetermined by the data but rather a more commonsensical observation about the situation scientists find themselves in. Values are a perfectly reasonable way to break such ties and do not commit us to deriving *is* from *ought*, for accepting a theory is not claiming it to be true. Roush argues, furthermore, that realist attempts to incorporate values without making them evidence for the truth of factual claims fail. Of course, others in this book will deny that constructive empiricism is defensible in the first place (compare Douglas's argument that values are implicated in the process of deciding the data). Nonetheless, those objections do not undermine Roush's claim that ties between theories are real and that scientists cannot reasonably be expected to have no favorites in such circumstances.

Doppelt has been a major contributor to debates about scientific rationality after Kuhn. His chapter applies views on these topics developed over a number of years to the ideal of value-free science. Doppelt argues that quick social constructivist arguments—group interests determine scientific outcomes—are simpleminded and suggests that a much more nuanced discussion of epistemic and nonepistemic values in rationality is called for. The account he provides makes two key claims in the context of this volume: (1) that epistemic values are contextual, local, and substantive theses about the world rather than logical and universal and (2) that nonepistemic interests are relevant to epistemic ones, but the latter do not reduce to the former. The first claim echoes Sober's point that evidential relevance is not a priori (though Doppelt probably wants to take this point further than Sober would) and directly denies the "egalitarian" view that Roberts believes logical empiricism is committed to. Abstract empirical virtues like simplicity don't do much work in real scientific debates; scientists give those virtues different concrete readings and thus differ over what counts as good evidence, explanation, and inference. Epistemic values are values, and debate over them has the character of any debate about values. Reasons can be given, the facts about the world matter, and simply wanting something to be true does not make it so. But the process is not driven by a universal fixed essence of rationality.

Kincaid takes up where Doppelt leaves off and applies and extends this general vision. Following recent work in epistemology (Williams 1996), he describes a general view labeled "contextualism." Contextualism denies that there are global criteria for deciding which beliefs or principles have epistemic priority. Kincaid argues that this claim has far-reaching and unappreciated implications for the ideal of value-free science. He looks at arguments both for and against value freedom and finds that they rest on precisely the premises about justification that contextualism rejects. In particular, Kincaid tries to show these assumptions at work in numerous arguments for the subjectivity of moral values, for it is this claim that motivates the value-free ideal. Using case studies from medicine and development economics, he concludes that it is true both that values have a rightful place and that they can have no place, because asking about the place of values rests on assumptions that are misguided from the start.

NOTES

1. The latter sense is much less contentious, though still of considerable interest.

2. These doubts are not universally taken to be decisive; variants of expressivist accounts of morality and a strong fact-value distinction still have sophisticated defenders.

REFERENCES

Ayer, A. 1952. *Language, Truth and Logic*. New York: Dover.

Blumer, H. 1971. "Social Problems as Collective Behavior." *Social Problems*, 18, pp. 298–306.

Duhem, P. 1991. *The Aim and Structure of Physical Theory*. Princeton, NJ: Princeton University Press.

Haack, S. 2003. "Knowledge and Propaganda: Reflections of an Old Feminist." In Pinnick et al. 2003.

Hume, D. 1888. *A Treatise of Human Nature*. Oxford: Clarendon.

Koertge, N. 2003. "Feminist Values and the Values of Science." In Pinnick et al. 2003.

Kuhn, T. 1962. *The Structure of Scientific Revolutions*. Chicago: University of Chicago Press.

Longino, H. 1990. *Science as Social Knowledge*. Princeton, NJ: Princeton University Press.

Longino, H. 2002. *The Fate of Knowledge*. Princeton, NJ: Princeton University Press.

Merton, R. K. 1966. "Social Problems and Sociological Theory." In R. K. Merton and R. A. Nisbet, eds., *Contemporary Social Problems*, 2nd ed. New York: Harcourt, Brace and World.

Myrdal, G. 1970. *Objectivity in Social Science*. London: Duckworth.

Pinnick, C., N. Koertge, and R. Almeder. 2003. *Scrutinizing Feminist Epistemology*. New Brunswick, NJ: Rutgers University Press.

Putnam, H. 2002. *The Collapse of the Fact-Value Dichotomy and Other Essays*. Cambridge, MA: Harvard University Press.

Quine, W., and S. Ullian. 1978. *The Web of Belief*. New York: McGraw-Hill.

Rudner, R. 1953. "The Scientist Qua Scientist Makes Value Judgments." *Philosophy of Science*, 20, pp. 1–6.

Solomon, M. 2002. *Social Empiricism*. Cambridge, MA: MIT Press.

Stevenson, C. 1960. *Ethics and Language*. New Haven, CT: Yale University Press.

Weber, M. 1968. *The Methodology of the Social Sciences*, tr. and ed. E. Shils and H. Finch. Glencoe, IL: Free Press.

Williams, M. 1996. *Unnatural Doubts*. Princeton, NJ: Princeton University Press.

Woolgar, S., and D. Pawluch. 1985. "Ontological Gerrymandering: The Anatomy of Social Problems Explanations." *Social Problems*, 32, pp. 214–227.

PART I

CASE STUDIES

ONE

FACT AND VALUE

John Dupré

THERE IS A VIEW OF SCIENCE, AS STEREOTYPED IN THE HANDS OF ITS
critics as its advocates, that goes as follows: Science deals only in facts.
Values come in only when decisions are made as to how the facts of science are to be applied. Often it is added that this second stage is no special concern of scientists, though this is an optional addition. My main
aim in this chapter is to see what sense can be made of the first part of
this story, that science deals only in facts.[1]

The expression "deals in" is intentionally vague. Two ways of dealing fairly obviously need to be considered. First is the question of the
nature of the products of science. These are certainly to be facts. But
there might also be a second question about inputs. In generating a fact,
say, dinosaurs are extinct, one needs to feed some facts in. (These are di-
nosaur bones. Our best tests suggest they are 80 million years old. No
dinosaurs have been observed recently. And so on.) So these inputs had
better be facts, too.

There are some obvious immediate worries. One might reasonably
object to the suggestion that the only products of science are facts with
the observation that science often produces things. Polio vaccines, mobile phones, laser-guided missiles, and suchlike are often thought of as
very much what science is in the business of producing. According to

the stereotypic view with which I began, it may be replied that science produces laws and suchlike, on the basis of which it is possible to create polio vaccines, mobile phones, and so on. And the trouble with this is that it seems grossly to misrepresent how science actually works. A group of scientists trying to develop a vaccine do not try first to formulate general rules of vaccine development and then hand them over to technicians who will produce the actual vaccines. No doubt they will benefit from the past experience, recorded in texts of various kinds, of past vaccine makers. And perhaps, if they are successful, they will themselves add to the body of advice for future vaccine makers. But it seems beyond dispute that the primary objective here is an effective vaccine, not any bit of fact or theory.

Let us ignore this concern for the time being, however, and concentrate on the question whether, insofar as science produces what we might think of as bits of discourse, these bits of discourse are strictly factual, never evaluative. So we need to ask what the criterion is for a bit of discourse being merely factual.

It is not hard to find some paradigm cases. "Electrons have negative charge" is pretty clearly factual, whereas "torturing children is a bad thing to do" is pretty clearly evaluative (though we might note at the outset that the clarity of this judgment strongly invites the suggestion that it is also a fact). The existence of these and many other possible paradigms may tempt one to apply the criterion famously applied to obscenity by U.S. Supreme Court Justice Stewart Potter, "I know one when I see one." But it is just as easy to find cases that are less clear. Consider, for instance, "The United States is a violent country." On the one hand, we can imagine a sociologist devising an objective measure of social violence—number of murders per capita, number of reported cases of domestic violence, and so on—and announcing that the United States ranked higher than most comparable countries in terms of this measure. But on the other hand, we can imagine someone describing this conclusion as a negative judgment.

Of course, there is a familiar response here. We have the fact and then the judgment. The fact is that there are certain statistics about acts of violence. The value judgment is that these statistics constitute a bad thing about the place where they were gathered. In support of this distinction, we can point out that it is always possible to accept the fact and reject the value judgment. Some people approve of violent countries (they reveal the rugged independence of the populace, perhaps), and perhaps there are even people who think torturing children is a good thing. But this defense is beside the present point. That point was just that the statement "The United States is a violent country" cannot be obviously assigned to either of the categories, factual or evaluative. In case this is not clear, compare the statement "Sam is a violent little boy." In any normal parlance, this does not mean just that Sam is disposed to occasional violent acts—that is, after all, true of virtually all little boys—still less that his rate of violent act production reaches a certain level on a standard scale approved by the American Psychological Association. It is a criticism of Sam, and probably of his parents, too. Anyone who doubts this should visit their nearest day care center and try out this comment on the parents collecting their precious charges there.

Suppose, as I have imagined with the case of social violence, that there is indeed a standard measure of violence for little boys. On this scale, a violent child is defined as one who emits more than five acts of aggression per hour. Now when I, as an expert child psychologist, announce that Sam is a violent child, my remark is entirely factual. Should his parents find the remark objectionable, I shall point out that this is no more than a factual observation, and it is entirely a subjective opinion, and one that I as a scientist shall certainly refrain from entertaining, whether it is a bad thing to be a violent child.

A possible conclusion at this point would be something like this: "The United States is a violent country" and "Sam is a violent little boy" are both potentially ambiguous. Although both may often be used evaluatively, especially by regular folk, scientists use them only after careful definition (operationalization) of their meanings. Thus, when used by responsible scientists, these statements will turn out to be merely and wholly factual. The statements under consideration are thus seriously ambiguous.

So perhaps scientists would do better to avoid these normatively loaded terms and stick to an explicitly technical language. To say that Sam scored 84 on the Smith-Jones physical assertiveness scale is much less threatening (even if this is practically off the scale, the sort of score achieved by only the most appallingly violent children). And it is certainly true that psychologists or psychiatrists, to pursue the present example, are often more inclined to invoke technical diagnostic language, backed up by detailed technical definitions in standard nosological manuals, than to say, for instance, that someone is mad.

There is, however, an overwhelming advantage to ordinary evaluative language: It provides reasons for action. To say that the United States is a violent country is a reason for politicians to act to reduce violence or mitigate its effects (for example, by controlling the availability of dangerous weapons). It is, other things being equal, a reason not to live there. And so on. It is of no interest just to be given a number and told this is the violence index for a country or a city; we want to know whether it is high or low or, indeed, whether it is good or bad. Similarly, though here we tread on shakier ground, it might be valuable to know that someone is mad. It might be expedient to restrain them, or at least not put them in charge of security at the local nuclear power station.

There is a general point here. Once we move away from the rarified environments of cosmology or particle physics, we are interested in scientific investigations that have consequences for action. And this undoubtedly is why, while often paying lip service to operationalized or technical concepts, scientific language often gets expressed in everyday evaluative language.

The situation so far seems to me to be this: Many terms of ordinary language are both descriptive and evaluative. The reason for this is obvious. Evaluative language expresses our interests, which, unsurprisingly, are things we are interested in expressing. When we describe things, it is often, perhaps usually, in terms that relate to the relevance of things for satisfying our interests. Sometimes we try to lay down rather precise criteria for applying interest-relative terminology to things. These range from the relatively banal—the standards that must be met to count as a class 1 potato, for instance—to the much more portentous, the standards that an act must meet to count as a murder. In such cases, we might be tempted to say that the precision of the criteria converts an evaluative term to a descriptive one. It is important to notice, however, that the precision is given point by the interest in evaluation. The same is often the case for operationalized terms in science. More often in everyday life, the terms are a much more indeterminate mix of the evaluative and the descriptive: crisp, soggy; fresh, stale, or rotten; vivacious, lethargic, idle, stupid, or intelligent; or, recalling Austin's memorable proposal for revitalizing aesthetics, dainty and dumpy.

This, I think, is the language that we use to talk about the things that matter to us, and to understand such language requires that we understand both the descriptive criteria and the normative significance of the concepts involved. It seems to follow that there is no possibility of drawing a sharp fact-value distinction. Science may reasonably eschew some of these familiar terms on the ground that they are vague and imprecise and may try to substitute more precisely defined alternatives. But first, the use of these alternatives will ultimately depend on their capturing the evaluative force of the vaguer terms they replace. And second, science does not, and almost certainly cannot, entirely dispense with the hybrid language of description and evaluation. This fact makes the assumption of a sharp fact-value distinction not only untenable but also often harmful.

So much for the general background of skepticism about the fact-value distinction. For the rest of this chapter, I shall be concerned with more detailed specific examples. Two such examples will illustrate more concretely how normativity finds its way into scientific work and how its denial can potentially be dangerous.

Before continuing, though, I have one more very general comment. The examples that I shall discuss will both be drawn from parts of science directly connected to human concerns. I have often heard the view expressed that though it is interesting and important that the human sciences should be contaminated with values, it is not altogether surprising. But what would really concern the advocate of the value-neutrality thesis with which this chapter began would be an indication that physics or chemistry or mathematics was value laden. So, on such a view, I am dodging the really important task.

In reply, let me first say that I do not propose to deny that many of the results of these sciences may well be value free. The sense in which I am questioning the legitimacy of the fact-value distinction is not one that implies that there are no areas that human values do not infiltrate. It is rather that there are large areas, including the domain of much of science, in which the attempt to separate the factual from the normative is futile. What I want to say about physics is that if most or all of physics is value free, it is not because physics is science but because most of physics simply doesn't matter to us. Whether electrons have a positive or a negative charge and whether there is a black hole in the middle of our galaxy are questions of absolutely no immediate importance to us. The only human interests they touch (and these they may indeed touch deeply) are cognitive ones, and so the only values that they implicate are cognitive values. The statement that electrons have negative charge is thus value free in a quite banal sense: It has no bearing on anything we care about.

I said that these were matters of no immediate importance, and the word *immediate* is crucial. It is often pointed out that physics also tells us how to build nuclear power stations and hydrogen bombs. Here, we are, to say the least, in the realm of values. There is no unique nuclear power station that physics tells us how to build, nor could there be a general theory that applied to the building of any possible power station. Physics assists us in building particular kinds of power stations, and particular kinds of power stations are more or less safe, efficient, ugly, and so on. Anyone who supposes there is a value-free theory of nuclear power station building, let alone hydrogen bomb construction, is, it seems to me, a fool or a liar. The argument that physics is value laden beyond the merely cognitive values mentioned in the last paragraph seems most plausibly to depend on some such claim as that physics really is, contrary to appearances or propaganda, the science of bomb building. I make no judgment on this issue. My point today is just that the value freedom of physics, if such there be, has no tendency to show that science is in general value free.

1.1 Rape

My first example is not a pleasant one. It is the evolutionary psychological hypothesis about rape.[2] The basic story goes something like this: In the Stone Age, when the central features of human nature are said to have evolved, females were attracted to mates who had command of resources that could be expended on rearing children. Perhaps they were also attracted to males with good genes—and perhaps these were simply genes for being, in the virtuously circular sense characteristic of sexual selection, attractive. Perhaps these ancestral females were smart enough to deploy some deception on the resource-rich males and get their resources from the "Dads" and their genes from the more attractive "cads." At any rate, there would very probably have been males with neither competitive-looking genes nor resources, and they, like everyone else, would be looking for a sexual strategy. Because they have no chance of persuading any females to engage in consensual sex with them, this strategy can be only rape. As is generally the way with evolutionary psychology, once a form of behavior has been proposed as a good idea in the Stone Age, it is inferred that a module for producing it must have evolved. So men, it appears, have a rape module, activated when they find their ability to attract females by any acceptable method falls to a low enough level.

Evolutionary psychologists presenting such theories generally also insist on a quite naive version of the fact-value distinction. Their claimed discoveries about rape are merely facts about human behavior, certainly not facts with any sort of evaluative consequences. We can at least agree, contrary to what evolutionary psychologists sometimes accuse their critics of maintaining, that showing that rape is, in the sense just described, natural doesn't mean it is good. Earthquakes and the AIDS virus are, discounting some paranoid speculations, natural but not thereby good. But such theories certainly do have consequences for what would be appropriate policy responses to the incidence of rape. Even this indisputable fact is enough to refute the occasional claim that such theories have no evaluative consequences. They have at least the consequences that certain policies would be good or bad. The most obvious such policy response to the theory in question would be the elimination of poverty, since the hypothesis is that it is poor men who are rapists (because they lack the resources to attract women). Though certainly a good idea, this goal has unfortunately proved difficult to achieve. On some plausible Marxist analyses, it is a goal that could not be achieved without the elimination of capitalism — an equally tricky proposition — because, on these analyses, poverty is not an intrinsic property of people but a relation between people, and a relation that is fundamental to capitalism. And it is interesting that such an analysis appears relevant to the sociobiological stories: It is not the intrinsic worthlessness of the failed caveman that doomed him to sterility or sexual violence, but his relative lack of worth compared with his more fortunate rivals.

But all of this is, of course, somewhat beside the point. Those who have thought seriously about contemporary sexual violence as opposed to the hypothetical reproductive strategies of imagined ancestors have observed that rape is not exclusively, or even mainly, a crime of resourceless reproductive predators lurking in dark alleyways but has much more to do with misogyny, and more to do with violence than sex, let alone reproduction. Its causes appear, therefore, to be at the level of ideology rather than economics.

These implications indicate that the stakes are high in theorizing about matters of this moment, but they do not get to the heart of my present argument. So far, I have spoken as if there is no problem whatever in deciding what, in the context of this theoretical inquiry, we are talking about. Indeed, to make research simpler, sociobiologists often begin their investigation of rape with observations of flies or ducks. If we have a good understanding of why sexually frustrated mallards leap out from behind bushes and have their way with unwilling, happily partnered, passing ducks, then the essential nature of rape is revealed, and we can start applying these insights to humans. Of course, what this blatantly ignores is the fact that human rape (and I doubt whether there is any other kind) is about as thoroughly normative a concept as one could possibly find. Those who supposed they were investigating the causes of rape but, since they were good scientists, were doing so with no preconceptions as to whether it was a good or a bad thing, are deeply confused: They lack any grasp of what it is that they are purporting to investigate.

All this is perfectly obvious when one looks at real issues rather than pseudoscience. A more serious perspective on rape is that it involves a profound violation of the rights of its victims. When, not long ago, it was conceptually impossible for a married man to rape his wife, this reflected a widespread moral assumption that, vis-à-vis her husband, a woman had no rights. Indeed, the husband was supposed to have a right, perhaps divinely guaranteed, to whatever kinds of sexual relations he desired with his wife. Nowadays, more complex debates surround the concept of date rape, the exact tones of voice in which no means yes, and so on. Less controversially, it has long been understood that sexual relations with young children is a form of rape, because the relation between adults and small children does not permit meaningful consent. But the age at which consent becomes possible varies greatly from culture to culture and is often subject to renegotiation.

The point of this is not to argue that there is no place for science in relation to such a topic. On the contrary, there are quantitative and qualitative sociological questions, psychological questions, criminological questions, and no doubt others that are of obvious importance. The point is just that if one supposes one is investigating a natural kind with a timeless essence, an essence that may be discovered in ducks and flies as much as in humans, one is unlikely to come up with any meaningful results. Though this is an extreme example, in that the value ladenness in this case is so blindingly obvious that only the most extreme scientism can conceal it, I think it is atypical only in that obviousness. As I argued in the opening section of this chapter, fact and value are typically inextricably linked in the matters that concern us, and we are most often concerned with matters that concern us.

1.2 Economics

My second example is a quite different one. Nowhere is the tradition of dividing the factual from the evaluative more deeply ingrained than in economics. In recognition of the fact that issues about the production and distribution of the goods on which human life depends do have a normative component, there is, indeed, a branch of economics called normative, or welfare, economics. But this is sharply divided from the properly factual investigations of so-called positive economics, and it is hardly a matter of debate that it is the latter that is the more prestigious branch of the discipline. In common with traditional positivism and contemporary scientism, the underlying assumption of this distinction is that there is a set of economic facts and laws that economists are employed to discover and that what to do with these is largely a matter for politicians or voters to decide.[3]

And in fact, normative economics has itself tended to reinforce this perspective and therefore tried to limit itself to the question whether there are economic actions that are indisputably beneficial. This concern is expressed in the focus of attention on the criterion of Pareto optimality: An economic allocation is said to be Pareto optimal if there is no possible transfer of goods that would improve the lot of some agent or agents while harming no one. It may be that failures to achieve Pareto optimality should be addressed where possible (though even this may be called into question by some accounts of distributive justice). But the "optimality" in "Pareto optimality" is a dubious one. If, for example, I possess everything in the world and I derive pleasure from the knowledge that I own everything in the world, this distribution of goods constitutes a Pareto optimum. If some crust of my bread were diverted to a starving child, I would no longer have the satisfaction of owning everything in the world, and similarly with any other possible transfer. So one person, myself, would be less well off. But this would be an unconvincing argument that this distribution was optimal, or even good. There are, of course, countless Pareto optima, which by itself suggests something anomalous in the use of the term *optimum*.

The problem is perfectly obvious. Although we can all agree that Pareto optimality is a good thing if we can get it, the issue of interest is which of the many Pareto optima we should prefer. Pareto optimality is really about efficiency, whereas we are interested in properly normative economics in matters such as justice. We should recall here the general assumption that science in general, and economics in particular, should aim simply to describe the mechanisms of economic activity and leave it to others to decide what to do with it. Not only is this assumption at work in positive economics but also it is even more starkly visible in much of the practice of normative economics, which is concerned not with how economies ought to be organized but with efficiency.

I believe that this is a highly undesirable, and very probably incoherent, conception of the business of economists. One way to see that it is undesirable is to note that when we consult supposedly expert economists about what might be good economic policy, we might naively suppose that they would have useful advice to offer us. But on the conception under review, it turns out that, apart perhaps from pointing to the occasional departure from Pareto optimality, they have no relevant expertise whatever. They are, after all, experts in efficiency, not policy. But because economists often seem willing to offer such advice, it seems disingenuous that they should deny that normative questions are part of their discipline. And if they do insist on this denial, they will presumably be of much less use to us than we had thought, and we could perhaps get by with rather fewer of them.

More worrying, it is quite clear that there is an implicit normative agenda to the vast majority of economic thinking. Because economists believe they have something to say about economic efficiency, they are naturally inclined to think of this as a good thing. And as the clearest measure of efficiency is the ability to produce more stuff with the same resources, economists are often inclined to think the goal of economic activity is to produce as much stuff as possible. Even if this account of the etiology of this goal is disputable, it is hard to dispute that many economists do assume such a goal, and assuming a goal is a good way of avoiding the vital intellectual labor of considering what the goals of economic activity really should be. Returning to the economists who offer advice on matters of public policy, I note that very frequently they assume that what they are required to do is advocate those policies that they believe, rightly or wrongly, will promote the production of as much stuff as possible.

In fact, even if we agree that something should be maximized by economic activity, an enormously difficult question is what that something should be. Not infrequently, positive economics assumes that the real question is about maximizing wealth measured in monetary terms, and tragically, many politicians seem willing to accept this facile view. An obviously preferable goal would be something like standard of living, except that would be little more than a marker for the difficult question of what constitutes standard of living. The work particularly of Amartya Sen[4] has made it clear that any satisfactory analysis of this concept will be only marginally related either to any standard account of utility or to the accumulation of wealth. It is also clear that even if we knew what constituted standard of living, we would still have to face the task of deciding how this should be distributed. Surely, the utility of increases in standard of living declines as one reaches more comfortable levels, so greater good can be gained by distributing standards of living more equally. And there is also the question of who should be among the beneficiaries of a distribution. Should we care about the standards of living of foreigners, for instance? Do the as yet unborn have any claim on a decent standard of living? Must we consider the well-being of non-human animals or the effects of economic activity on the environment?

Once again, however, the issue I want to emphasize here is the inescapably value-laden nature of the terms in which we talk about ourselves and our social existence. Consider a central idea in macroeconomics, the measurement of which has had profound implications on economic policies throughout the world, inflation. Like earthquakes or AIDS, inflation is generally seen to be a bad thing. But also like earthquakes and AIDS, it is seen as the sort of thing that can be described and theorized without regard to its goodness or badness.

The problem here is somewhat different from that for rape. The normative judgment is fundamental to the meaning of rape and therefore fundamental to negotiations about what should and should not count as rape. With inflation, normativity comes in a little later. The primary problem, as has long been familiar to economists, though it often appears to surprise others, is that there is no unequivocal way of measuring this economic property. It would be easy enough if everything changed in price by identical percentages, but of course that does not happen. How should we balance a rise in the price of staple foods, say, against a fall in the price of air travel? The immediately obvious reply is that we should weight different items in proportion to the amount spent on them. The problem, then, is that not all goods are equally consumed by all people or even by all groups of people. It is quite commonly the case that luxury goods fall in price while basic necessities rise. It might be that these cancel out under the suggested weighting, so that there is no measured inflation. But for those too poor to afford luxury goods, there has manifestly been an increase in the price level.

How, then, does one decide how such an index should be constructed? The unavoidable answer, it seems to me, is that it depends on the purposes for which it is to be constructed. There are many very practical such purposes. People on pensions, for instance, may have their incomes adjusted to account for changes in the level of inflation. For such purposes, the goal might reasonably be to maintain the value of the pension, in which case the ideal would be to enable typical pensioners to continue to afford the goods that they had previously consumed. Of course, no pensioner is absolutely typical, but a case might be made for addressing particularly the case of pensioners dependent solely on the pension. For such ends, it would clearly be desirable to have specific indices designed for specific groups. But the goals might be quite different, calling for different measures. For example, and perhaps more plausibly, one such goal might be to save the taxpayer money.

Perhaps the central goal nowadays of inflation measurement is as an input into the decision procedures of central banks in determining interest rates. In Britain (I'm not sure how widespread the practice is), this leads to the rather bizarre habit of regularly announcing something called the "underlying rate of inflation." This is a measure of inflation that ignores changes in mortgage payments consequent on changes in interest rates. The rationale for this appears to be that the article of faith on which much macroeconomic policy depends is that the inflation rate is inversely related to interest rates. Since increasing interest rates has an immediate and large effect in increasing the prices confronted by consumers, this central dogma would be constantly refuted if mortgage costs were included in the measure of inflation. Hence the underlying rate is important as a way of allowing the theory to be maintained.

(I suppose this aspect of the matter is of more obvious concern to students of the theory laden than of the value laden.)

Yet another aspect of all this is that the assumption that inflation is objectively bad is by no means simple. In common with most middle-class Americans, I have spent substantial parts of my life owing large sums of money borrowed at fixed interest rates. From a personal point of view, therefore, I have always seen inflation as something to be enthusiastically welcomed. The deep horror with which it is now perceived should lend support to those who believe that the world is mainly controlled by bankers.

Some quite different aspects of value ladenness could be introduced by considering another central macroeconomic concept, employment. Having work is widely perceived in many contemporary cultures as a necessary condition for any social status and even for self-respect. But what counts as work is a complicated and contentious issue and one that has profound implications for all kinds of economic policies. It is still frequently the case, for instance, that work is equated with the receipt of financial reward, with the consequence that domestic work, from raising children to the domestic production of food, was, from an economic perspective, a form of unemployment. A quite different concept can be found in Adam Smith (and an earlier Adam who was required to make his living "in the sweat of thy face"), in which work is generally unpleasant— toil and trouble—and understood by its contrast to leisure or ease (see Smith 1994, 33). Quite different again is the idea, most conspicuously developed by Karl Marx, that work provides the possibility of human self-fulfillment. Both these conceptions are evidently value laden, and the notion that there can be a purified economic conception of work, somehow divorced from any of these varied normative connotations, seems both misguided and potentially dangerous.[5] There are, in sum, many ways in which values figure in the construction and use of many of our concepts, and scientific concepts are no exceptions. For much of language, the notion of separating the one from the other is altogether infeasible.[6]

1.3 Conclusion

As I indicated earlier, I am not claiming that there is no distinction between the factual and the normative. What I do claim is that this is not a distinction that can be read off from a mere inspection of the words in a sentence or a distinction on one side or the other of which every concept can be unequivocally placed. For large tracts of language—centrally, the language we use to describe ourselves and our societies—the factual and the normative are thoroughly interconnected. Where matters of importance to our lives are at stake, the language we use has more or less profound consequences, and our evaluation of those consequences is deeply embedded in the construction of our concepts. The fundamental distinction at work here is that between what matters to us and what doesn't. There are plenty of more or less wholly value-free statements, but they achieve that status by restricting themselves to things that are of merely academic interest to us. This is one reason that physics has been a sometimes disastrous model for the rest of science. We hardly want to limit science to the investigation of things that don't matter much to us one way or the other. The application of assumptions appropriate only to things that don't matter to those that do is potentially a disastrous one.

NOTES

1. I am grateful to Francesco Guala and Harold Kincaid for helpful comments on earlier versions of this chapter. This work was completed as part of the program of the Economic and Social Research Council (ESRC) Centre for Genomics in Society (Egenis). The support of the ESRC is gratefully acknowledged.

2. A standard reference is Thornhill and Thornhill (1992). The ideas were popularized by Thornhill and Palmer (2000). For detailed rebuttal, see various essays in Travis (2003).

3. A classic paper by Friedman (1953) provides a well-known statement of this position.

4. A number of insightful discussions of the issue can be found in Nussbaum and Sen (1993).

5. These different meanings of work are discussed in more detail in Dupré (2001, 138–46) and Gagnier and Dupré (1995).

6. For more detailed accounts of important aspects of value ladenness in economics, see Starmer (2000) and Guala (2000).

REFERENCES

Dupré, J. 2001. *Human Nature and the Limits of Science.* Oxford: Oxford University Press.
Friedman, M. 1953. "The Methodology of Positive Economics," *Essays in Positive Economics*, pp. 3–43. Chicago: University of Chicago Press.

Gagnier, R., and J. Dupré. 1995. "On Work and Idleness." *Feminist Economics*, 1, pp. 1–14.

Guala, Francesco. 2000. "The Logic of Normative Falsification: Rationality and Experiments in Decision Theory." *Journal of Economic Methodology*, 7, pp. 59–93.

Nussbaum, M., and A. Sen, eds. 1993. *The Quality of Life*. Oxford: Oxford University Press.

Smith, A. 1776/1994. *The Wealth of Nations*. Ed. E. Cannan. New York: Modern Library.

Starmer, Chris. 2000. "Developments in Non-Expected Utility: The Hunt for a Descriptive Theory of Choice under Risk." *Journal of Economic Literature*, 38, pp. 332–82.

Thornhill, R., and C. T. Palmer. 2000. *A Natural History of Rape: Biological Bases of Sexual Coercion*. Cambridge, MA: MIT Press.

Thornhill, R., and N. W. Thornhill. 1992. "The Evolutionary Psychology of Men's Coercive Sexuality." *Behavioral and Brain Sciences*, 15, pp. 363–421.

Travis. C. B., ed. 2003. *Evolution, Gender, and Rape*. Cambridge, MA: MIT Press.

TWO

SOCIAL PROBLEMS
Michael Root

2.1 Introduction

Should a sociologist, when studying child abuse or homelessness, say that child abuse or homelessness is wrong? Max Weber thought not. At the 1909 meetings of the Social Policy Association, he argued that sociology should be a science and a science must be silent on questions of right and wrong (Proctor 1991). Weber's view continues to influence sociology, and his understanding of science has shaped the sociological study of social problems and deviance.[1]

Studies of juvenile delinquency, mental illness, alcoholism, child abuse, divorce, illegitimacy, poverty, and homelessness came to prominence within sociology in the 1950s with the founding of the Society for the Study of Social Problems (SSSP) and the publication of the journal *Social Problems* (Rose 1971).[2] Most of the sociologists who founded the society and began the journal shared at least two interests: (1) to make sociology more socially relevant and (2) to make studies of social problems objective and their findings scientific.[3] The members assumed, as Weber had, that (2) requires that their studies be value free and, in particular, free from their own expression of right and wrong.[4]

Robert K. Merton was not a member of the SSSP but was committed to both (1) and (2) (Merton 1966). He wrote often about social problems and believed that his own statements about juvenile delinquency and alcoholism were value free because he never said that delinquency and alcoholism were bad or should be disapproved of; though he referred to juvenile delinquency and alcoholism as problems, Merton meant that they were dysfunctional but not that they were wrong.

In 1971, Herbert Blumer, a prominent member of the SSSP, published a paper critical of Merton's way of doing sociology, but he agreed that a sociological study of social problems should be value free (Blumer 1971).[5] A study of alcoholism, according to Blumer, should describe how the public responds to heavy drinking and be silent on whether the practice is right or wrong. The disagreement between Blumer and Merton was over how and not whether to keep the study of alcoholism or delinquency free of the sociologist's own moral values or expressions of disapproval.[6]

The first section of the chapter describes the history of the sociological study of social problems. The second explains how the history has been shaped by Weber's views on value freedom. The third shows why a study of social problems cannot be silent on questions of right and wrong and, at the same time, be socially relevant, and the fourth section considers whether a study can be both value laden and objective or scientific.

2.2 Merton and Blumer

Some parents in the United States beat their children. The question for the sociologist is whether to call the practice a social problem. On Merton's view, the answer depends on whether the practice is dysfunctional. Practices are dysfunctional, according to Merton, when they unsettle a community by violating norms members are expected to abide by. Child beating is a social problem in the United States, on Merton's view, because the practice violates a norm (accepted by most Americans) for how children ought to be treated.

"Norms," according to Merton, "may prescribe behavior or proscribe it; or they may merely indicate which forms of behavior are preferred or simply permitted" (Merton 1966, 817). Moreover, norms change over time and vary between communities. As a result, the beating of children might be a problem now but might not have been many years ago, or it might be a problem in the United States but not elsewhere. In addition, on Merton's view, a norm can be either latent or manifest, and, as a result, a practice can be a problem even if many members of a community do not think so. That is, a practice can have dysfunctional effects even if members are not aware of them or it unsettles a community's norms without a member's knowing so.[7]

Blumer opposed talk of hidden problems.[8] Problems, according to Blumer, rely on the public's disapproval.[9] A practice becomes a problem in the process of being contested and labeled. Alcoholism is a problem because the public disapproves of heavy drinking; they do not disapprove of heavy drinking because the practice is a problem.[10] Social problems, Blumer writes, are "products of a process of collective definition instead of existing independently as a set of objective social arrangements with an institutional makeup" (Blumer 1971, 298). They become problems when they enter the public's consciousness and come to be talked about in a particular way. As a result, a sociologist can decide whether alcoholism is a social problem only by observing how the public responds to heavy drinking.

Blumer's article was followed by a series of studies that assume, as Blumer does, that to be scientific, sociologists cannot label a practice a "problem" unless their subjects do. Here are a few examples. In an article entitled "The 'Discovery' of Child Abuse," Stephen Pfohl argued that though the abuse of children has a long history, the term *child abuse* and widespread opposition to the practice are recent (Pfohl 1977). The mistreatment of children has only recently become a matter of public concern, and the task of the sociologist is to explain how the concern arose and what social activities are helping to sustain it.

According to Pfohl, harsh treatment of children became a problem here because medical groups began campaigns to have the treatment labeled "abusive." At first, the public was not moved, but in 1960 pediatricians and psychiatrists joined forces to promote the label "battered child syndrome," and the new label won the public's attention. The doctors in Pfohl's account did not discover that child abuse was a problem but rather made the practice a problem by inventing the label and spurring the public's disapproval of harsh treatments of children. The task of the sociologist, on Pfohl's view, is to describe how the public came to disapprove rather than consider whether their disapproval is appropriate.

As a second example, stalking became a national concern in the 1990s, and by 1993 forty-eight states had passed antistalking laws, but before these laws, few talked about stalking at all. In 1995, Kathleen Lowney and Joel Best published a chapter on the sudden concern over stalking (Lowney and Best 1995). Their interest was not why people stalk or whether stalking is wrong but how a single label, "stalking," came to be applied to a variety of very different incidents and how the incidents became a topic of public policy. The cause, they explained, was the media.[11] The media did not cause people to stalk but packaged the behavior and made stalking into a kind, a series of incidents with a shared nature.

According to Lowney and Best, people came to see stalking where stalking would never have been seen before.[12] The media, on their view, did not discover stalking or that stalking was a social problem but invented stalking by applying a single label to a ragbag of events and encouraging the public to see them as the same. In writing about stalking, Lowney and Best talk about the public's unease over the practice but are silent on whether stalking should be disapproved of or whether the public's unease is reasonable.

As a third example, one third of all births in the United States are to unmarried mothers—a ratio that has been nearly constant since 1994 but that increased rapidly from the 1960s through the 1980s. According to the National Center of Health Statistics, out-of-wedlock births accounted for 5.3% of the nation's newborns in 1960, 10.7% in 1970, 18.4% in 1980, and 28% in 1990. The increase in percentage of out-of-wedlock births between 1960 and 1990 does not mean, on Blumer's view, that out-of-wedlock births were more of a problem in the United States in 1990 than in 1960; they were more of a problem, on his view, only if Americans showed greater concern over out-of-wedlock births in 1990 than in 1960.[13] Out-of-wedlock births, on Blumer's view, could have been less of a problem in 1990 than in 1960, even though out-of-wedlock births increased from 5.3 to 28%.

In 1996, the illegitimacy rate was declining, but the debate over "family values" and welfare dependency had reached a peak, and Congress funded a $400 million "illegitimacy bonus" program. The program offered cash awards to states that showed the largest decrease in out-of-wedlock births with no increase in abortion rates (Lewin 2000). On Blumer's view, illegitimacy was a much greater problem in 1996 than before (despite the drop in rate), because, as the bonus shows, the concern of Congress and the voting public had increased.

As a sociologist uses the term *social problem*, she can call out-of-wedlock births a social problem and remain silent on whether they are wrong. She can describe how Americans disapprove of having children outside of marriage without saying whether she does. However, because her use of the term is value free, if she finds that out-of-wedlock births are a social problem in the United States, her findings will have limited social relevance, since they say nothing about whether Americans should disapprove of the practice, which is the question many Americans are most concerned about.

2.3 Hidden Problems

The public, on Blumer's view, cannot be mistaken about whether alcoholism or illegitimacy is a problem; a condition is a problem within a population if and only if the members think so. Problems, for Merton, on the other hand, can be hidden from the members' view. Any connection between a problem and the public's disapproval, for Merton, is contingent, while, for Blumer, necessary. If x is a problem in S, then, according to Merton, members of S would disapprove of x if they understood that x is dysfunctional in S, but they might not know whether x unsettles a norm of S and so might not manifest any disapproval of the practice. If a sociologist says in 1960 that illegitimacy is a problem in the United States, he is describing how Americans would have responded in 1960 if they had understood the effects of out-of-wedlock births on American families and seen that many of the effects were dysfunctional. On Blumer's view, on the other hand, out-of-wedlock births were a problem in 1960 only if many Americans in 1960 expressed disapproval of them.

Blumer and Merton try to give terms like *social problem* and *deviance* a purely descriptive meaning. In ordinary language, when a speaker says that divorce is a social problem or deviant, she expresses her disapproval, but when Blumer or Merton uses the terms, he means to describe the public's attitudes rather than express his own. When Merton says that alcoholism is a social problem in S, he does not mean to express his own attitudes toward heavy drinking but only to report others' (manifest or latent) disapproval.[14] When Merton uses words like *deviant* or *dysfunctional*, he does not mean to say what ought to be but only what is within the community to which his subjects belong. "As used by the sociologist, the term deviant," Merton writes, "is a technical rather than moral one. It does not signify moral disapproval by the sociologist."[15]

On Merton's view, whether a dysfunctional condition is good or bad depends on a further and entirely independent judgment of the moral worth of the system in which the condition occurs. Within a caste system, for example, to offer the lower castes educational opportunities would be dysfunctional and thus a problem for the persistence of the castes. Offering the opportunities to a lower caste, though dysfunctional, is wrong, however, only if castes ought to be approved of. But on Merton's view, a sociologist is not able to approve or disapprove of castes and, at the same time, keep his studies value free, and as a result, he must remain silent on whether a caste system is good or bad and whether giving members of a lower caste educational opportunities ought to be approved or disapproved of.

Blumer and Merton agree that sociology cannot be a science if sociologists say what ought to be rather than what is. They share Max Weber's view that "there are no (rational or empirical) objective procedures of any kind whatsoever which can provide us with a decision [about what we should desire]" (Weber 1968, 18–19). Weber said that "it is simply naive to believe . . . that it is possible to establish and to demonstrate as scientifically valid 'a principle' for practical social science from which the norms for the solution of practical problems can be unambiguously derived" (Weber 1968, 56). According to Weber, a sociologist should try to keep her own values to herself when she studies a social problem and limit herself to a description of her subjects' values or how a practice does or does not violate them.

Blumer asserted that, by simply describing the public's concerns — their complaints and demands, the lawsuits, the press conferences, the letters of protest, the pickets or boycotts, the resolutions, and the media attention — and describing the forces behind them, the sociologist assures that his studies are scientific (an objective description of his subject's subjective judgments); he records the public's attitude toward a practice rather than exhibits his own. As long as the sociologist simply describes the public's sense of right and wrong and does not express his own, his study of the public's problems can be socially relevant and his findings scientific.

2.4 Labeling Theory

In 1985, Steven Woolgar published an article critical of Blumer's work and, in particular, of Blumer's claim that social problems are the result of a process of definition in which a given condition is picked out and identified as a problem (Woolgar and Pawluch 1985).[16] Woolgar agreed that the public makes x a problem by labeling x a problem but claimed, in addition, that the public makes x a condition by giving x a label. Conditions, according to Woolgar, are no more objective features of social life than problems are. Both are talked into being rather than "real."[17] On Woolgar's view, the sociologist does not discover alcoholism or divorce in nature any more than she discovers social problems or deviance there; alcoholism is a "socio-historical accomplishment," a subjective rather than an ontologically objective category just as problems are.[18]

The public invents alcoholism, according to Woolgar, by drawing a boundary around episodes of heavy drinking and making them into a thing (a single kind of conduct), just as the public invents social problems by drawing a boundary around a number of events and labeling them a problem. Thus, a sociologist should not say that alcoholism is a social problem in the United States but instead that Americans (1) make heavy drinking a condition by giving the events a single label and (2) make the events a problem by calling them "deviant" or labeling them "problems." By using (rather than mentioning) the label "alcoholism," the sociologist approves of the boundary her subjects have drawn around the events and, like her subjects, takes the events to be a condition; on Woolgar's view, to make her study of heavy drinking in the United States objective and her findings scientific, the sociologist should mention rather than use the label and not say that alcoholism is a condition but only that, according to her subjects, a condition and not say that heavy drinking is wrong but only a collection of events her subjects disapprove of.

Blumer died in 1986, but, in response to Woolgar's criticism, his supporters dropped the word *condition*.[19] Instead, they called alcoholism a "putative condition" and spoke of the claims-making activities of individuals or groups about "putative" rather than "real" conditions (Spector and Kitsuse 1987). Alcoholism only becomes a condition in S, they conceded, through the claims-making or labeling activities of individuals or groups in S. Before the members of S treated episodes of heavy drinking as tokens of a single type, as a kind, S had no alcoholism and no alcoholics for a sociologist to study or the public to grow concerned about. On Woolgar's view, to be objective, a sociologist should say how his subjects classify events but not whether "alcoholism" marks any real boundaries between them, much as, on Blumer's view, he should describe how his subjects use the term *social problem* but not say how they ought to.

2.5 Objectivity

A sociologist, according to Woolgar, should not use the label "alcoholism" or "problem" but only mention the terms or describe how her subjects use them; she should not say that in the United States alcoholism is a social problem or deviant but instead that Americans use the label "alcoholism" to unite (reify) a series of events (make them into an entity) and use the label "deviant" to make the entity into a problem. To be objective, she can count (1) how many Americans drink heavily but not (2) how many are alcoholic or how many are problem drinkers there. She can say (1) but not (2), for in saying (2) she would be expressing her own attitudes toward heavy drinking rather than simply describing the attitudes of her subjects.

1. The events that Americans call "alcoholism" and label "a problem" increased in the United States between 1960 and 1980.
2. Alcoholism became more of a problem in the United States between 1960 and 1980.

In saying (1), the sociologist reports how Americans see heavy drinking but keeps her own view of these events to herself. Because many sociological studies of social problems use rather than simply mention terms like *alcoholism* and *problems*, on Woolgar's view, many are not objective and their findings not scientific.

As an example, in 1976 some sociologists studied family violence in Cologne, Germany, and, in particular, violence against women there (Mies 1983). The behavior they wished to pick out and talk about was not at the time recognized by the residents as a condition, as a kind of conduct (what Woolgar calls an entity). That is, "wife battering" and "spousal abuse" were the sociologists' and not their subjects' categories. As a result, when the sociologists said that wife battering was a social problem in Cologne in 1976, they were not saying that the women of Cologne saw the events as a condition (entity) but saying how the events should be seen; the sociologists relied on their own categories to convince members of the community to disapprove of family violence. They said that wife battering is a condition and wrong rather than a putative condition and an apparent wrong.

A sociologist who says that x battered y (in contrast to saying that x struck y) implies that what x did to y is improper. He could mean that what x did is improper in the eyes of y, but that is not what the sociologists in Cologne meant, because in Cologne, apparently, the women did not see the striking as improper. When the sociologists in Cologne said that x battered y, they meant that, in their own eyes, what x did was improper. They invoked their own standard of impropriety, and, as a result, given Weber's way of thinking, they messed up; instead of sticking to the facts, they reached into the realm of politics or values.

Had the sociologists in Cologne listened to Herbert Blumer, they would have said that wife battering was not a problem there, because few in the city expressed any concern over the practice. Had they listened to Steven Woolgar, they would not have said that husbands battered their wives but only that they putatively battered them. But women in Cologne were lucky, for the sociologists did not listen to either Blumer or Woolgar but expressed their own disapproval of wife battering. Had their study been value free, their findings would not have been relevant to the interests of the women of Cologne and, in particular, to the interest of those at risk of being beaten by their husbands.

The sociologists in Cologne took a very different approach to the study of social problems than Kingsley Davis did when he studied homosexuality in the United States some years before (Davis 1966). In his study of homosexuality, Davis described how the practice offends the

American public but was silent on whether the public should be offended and whether homosexuality ought to be disapproved of. At the time he was conducting his study, however, Americans differed in their attitudes toward homosexuality, and even if many disapproved of the practice, some did not. In writing that homosexuality is a social problem, Davis relied on the attitudes of opponents and did not consider the attitudes of Americans who do not disapprove of the practice. He did not say, however, that homosexuality is a problem for the opponents of homosexuality but rather a problem in America. As a result, even if his study was value free and his findings scientific, Davis's study was not relevant to the debate within the United States over what one's attitude toward homosexuality ought to be. To address that debate, his study would have to describe not only what the opponents of the practice say but also whether they ought to be listened to.

2.6 Science and Values

The sociologist, Merton says, can describe her subject's values but not dispute them, discover their attitudes toward homosexuality but not say whether they are warranted. She can find out whether homosexuality is a problem in the United States, but she cannot say whether it ought to be; for her findings to be scientific, she has to keep them silent on whether homosexuality is wrong.[20] However, not every function, according to Merton, is manifest, and as a result, even if homosexuality had no manifest function within the United States, the practice could have been a latent one. But if homosexuality has a latent function, then the practice is not dysfunctional, even if many Americans disapprove of homosexuality. Moreover, if, as Merton maintains, x is a social problem in S if and only if x is dysfunctional in S, then, should homosexuality have a latent function, the practice would not be a social problem in the United States after all, and anyone in the United States who disapproves of homosexuality would be making a mistake. As a result, a sociologist, given Merton's view of social problems, should be able to say whether the moral attitudes of her subjects are warranted. She should be able to not only describe their attitudes but also say what they ought to be. She should be able say that in the United States many people take homosexuality to be a social problem but they ought not to, or that in the

United States homosexuality is not wrong, even though many people think so.

As long as a sociologist assumes, as Merton does, that a practice can have consequences members of a community are unaware of or that members can be unaware of whether the practice unsettles a norm, he is able to say whether their approval or disapproval of the practice is warranted. He can say that the members do not fully understand the effects of the practice, but if they did, they would see that they ought not disapprove of it, or he can say that the members do not understand the relevant norms, but if they did, they would see that the practice is not dysfunctional and, as a result, not wrong after all.

Unless a norm is only an expression of the public's taste, a sociologist is able to correct their statements of right and wrong. She can ask whether members of the community would choose to abide by the norm, were they to choose under conditions favorable to sound judgment.[21] Many current models of rational choice include descriptions of such conditions and, as a result, provide standards by which to assess a member's values.[22] The sociologist can ask whether her subjects would approve of wife battering if they were fully informed and reasonable, and given a reason to think they would not approve, she is able to object to their manifest approval of the practice.

In judging what subjects would endorse under ideal conditions, the sociologist is not replacing their values with his own, for, even if latent, the values are theirs (ones they would have endorsed had they been more reflective), but the sociologist's judgment is not entirely free of his own values either, for to decide what they would endorse, he has to rely on what, on his view, counts as full reflection and what, on his view, would survive his subjects' own critical review. If the sociologist (1) believes that the members of S are situated more or less as he is relative to a particular practice, (2) believes that in S norm N governs the practice, (3) believes that the members of S endorse N, and yet (4) cannot reflectively endorse N himself, then he has reason to believe that the endorsement of N by members of S is not fully reflective either.

Models of rational choice can be used in sociology to decide whether a society's values are reasonable, but for a sociologist to employ them, she will have to rely on her own sense of right and wrong; she cannot take herself or her own moral commitments entirely out of the picture. As a result, in saying that a value is latent (rather than manifest) in S, a sociologist is relying on her own sense of right and wrong and also the sense shared by members of S, and in saying that a practice is a manifest (apparent) but not a real problem in S, she is saying what is so in S, based on her own view of what ought to be.

Blumer assumes that values are subjective; he is an expressivist with respect to right and wrong. That is, for him, "x is wrong" means no more than that x is a practice of which I disapprove. Expressivism was common within sociology in the 1950s, when the Society for the Study of Social Problems was founded, but less common among sociologists today. In addition, whether a member of a community would express disapproval of alcoholism or heavy drinking were he fully informed and rational, many sociologists would now admit, is no less an objective or scientific question than whether they currently express disapproval of the practice.

Even if a sociologist can reasonably object to her subjects' values or concerns, why should she? Why are such assessments a proper task for sociology? The answer lies with the reasons why the Society for the Study of Social Problems and the journal *Social Problems* were started some fifty years ago. The sociology of social problems was meant to be socially relevant and, in particular, to diagnose and find cures for social ills. If all a sociological study of stalking or child abuse could show is how many members of a community disapprove of the practice or how the media influences a member's use of a label, the study would not be relevant to questions of policy, for before adopting a policy on stalking or child abuse, the public needs to know whether they ought to disapprove of the practice rather than whether they do or not.

2.7 Conclusion

Blumer and Merton wanted to make the study of social problems a science, but their work was inspired by the same interest, an interest in social reform, that led sociologists years before to study crime and delinquency in America's cities. They wanted their studies to be scientific but, in addition, inform debates over public policy and, in particular, over what should be or should not be socially sanctioned or permitted or forbidden by the nation's laws. But they had a false view of both science and values, and the limits they placed on their studies of social problems did not make their findings more scientific but did keep them from becoming more socially relevant.

NOTES

1. In this paper, I use the terms *deviance* and *social problem* interchangeably, but *deviance* is sometimes used more narrowly to cover practices that a community classifies as *criminal* (see, for example, Frank Tannenbaum 1951), and some sociologists use 'deviant' to cover any violation of any of the norms of a community and 'social problem' to cover a violation that the members disapprove of (see, for example, Tallman and McGee 1971).

2. An important early sociological study of social problems is Thomas and Znaniecki (1927), who studied the incidence of delinquency and crime among Polish immigrants in Chicago and explained how the shift from traditional forms of social control to the looser controls of modern urban life contributed to disorganization and conflict within a Chicago neighborhood.

3. Members of the SSSP wanted to replace a psychological or medical model with a more sociological study of deviance. Psychological or medical studies, in their view, too seldom question the label "deviant" or too often take the deviance of the acts for granted. Medical or psychological studies assume that some acts are inherently deviant and look for their causes in a person's head; according to the members of the SSSP, the important variable is what makes x deviant rather than what causes an individual to do x, and, as the members saw it, what makes x deviant is not any fact of nature but a social practice in virtue of which x is labeled "deviant."

4. Many assumed that to be scientific, a statement had to be objectively true or false and that statements of moral right or wrong are neither but mere expressions of preference or emotion.

5. Blumer does not mention Merton by name in this paper, but in offering his criticism of the functional approach to social problems, he seems to have Merton and his students in mind.

6. A few sociologists challenged the assumption that the sociological study of social problems could or should be free of the sociologists' own values. See, for example, Gross (1965). At least one president of the SSSP, Alvin Gouldner, maintained that studies of deviance could not be free of politics (Gouldner 1962, 199–213). Today there is more doubt within the SSSP than there was in the early years that sociologists should or can keep their own politics out of their studies of social problems, but as one member suggested recently, many still treat partisanship the way the Victorians treated sex: (1) they indulge and (2) they like it, but they have a habit of denying both (see Alvarez 2001).

7. On Merton's view, members of the community can fail to be aware that x is dysfunctional for either of two reasons: First, they can fail to recognize that y is a consequence of x, and second, they can fail to recognize that y opposes a community norm.

8. Talk of a social function, on Blumer's view, is unscientific. He writes (1971, 299–300): "Contrary to the pretensions of sociologists, sociological theory *by itself* has been conspicuously impotent to detect or identify social problems. This can be seen in the case of the three most prestigeful sociological concepts currently used to explain the emergence of social problems, namely the concepts of 'deviance,' 'dysfunction,' and 'structural strain.' These concepts are useless as means of identifying social problems. For one thing, none of them has a set of benchmarks that enable the scholar to identify in the empirical world the so-called instances of deviance, dysfunction, or structural strain." The emphasis is his.

9. Blumer writes (1971, 300): "A second deficiency of the conventional sociological approach is the assumption that a social problem exists basically in the form of an identifiable objective condition in society." By "objective condition," Blumer means a condition that the society need not be able to recognize.

10. Blumer's claim that a practice becomes a problem by being called a problem is anticipated in Howard Becker's *Outsiders: Studies in the Sociology of Deviance* (1963). Becker writes: "From this point of view, deviance is not a quality of the act the person commits, but rather a consequence of the application by others of rules and sanctions to an 'offender.' The deviant is one to whom that label has successfully been applied; deviant behavior is behavior that people so label" (p. 9). The view that labeling makes an act deviant appeared even earlier, in Edwin M. Lemert's *Social Pathology* (1951).

11. Lowney and Best write (1995, 48): "The media play various roles in social problems construction. Studies in the construction of crime problems suggest that the press sometimes serves as the primary claims maker (e.g., claims making about urban violence and freeway violence). Other problems emerge through secondary press reports of claims making by social activists (e.g., drunk driving, battering, rape), professionals (e.g., child abuse ad computer crimes) or criminal justice officials (e.g., serial murder)."

12. A recent reviewer of the 1950 film *All about Eve* writes: "We'd probably call Eve a stalker today. How else would you describe someone who attends every performance of her favorite actress's play and lurks by the stage door just to glimpse her passing by? In those more innocent times she was classified as an 'autograph fiend' " (Covert 2000).

13. According to Blumer, the increase in incidence might have prompted a greater concern, but the greater concern, rather than the rise in out-of-wedlock births, would be why the births are more of a problem in 1990 than in 1960.

14. In other words, Merton means that the effects of heavy drinking violate the norms of S and that, as a result, the members of S would disapprove of the practice, were they informed of the consequences.

15. Merton (1966, 805). Merton says the same about the term *dysfunctional*. He writes (p. 822): "Above all, it must be emphasized that the concept of a social dysfunction does not harbor an implied moral judgment. Social dysfunction is not a latter-day terminological substitute for immorality, unethical practice, or the socially undesirable. It is an objective concept, not a morally evaluating one. . . . Sociological analyses of function and dysfunction are in a different universe of discourse from that of moral judgments." He adds (p. 823): "The concept of social dysfunction is not based on ethical premises for it refers to how things work in society and not to their ethical worth."

16. Woolgar and Pawluch (1985, 214–27).

17. The term *real* means different things to different people, but the issue here is when kinds or categories are real, for example, when a category like "alcoholism" or "illegitimacy" is real rather than made-up or constructed by the public. As best as I can tell, a category C, on Woolgar's view, is real only if an x could belong to C, even if there were no S whose members sort by C, that is, no S whose members, for any x, believe that Cx or not Cx. Woolgar believes that if Americans did not classify births as legitimate or illegitimate, then there would be no legitimate or illegitimate births here, and as a result, Woolgar believes that "illegitimacy" is not real. For a discussion of different ways of understanding "real kind," see Root (2000).

18. See Searle (1995, 1–13) for a discussion of the distinction between ontologically objective and ontologically subjective categories. As Searle explains, a category C can be ontologically subjective, depend on our thought or talk, and be epistemically objective in the sense that given any x whether Cx or not Cx can be objectively determined. Categories like "divorce" are ontologically subjective, for there would be no divorce, were there no institution of marriage, but given the institution, whether a man is divorced is an epistemically objective matter. What makes it true that a person is divorced is an institution, but given the institution, we can objectively judge whether anyone is divorced or not. On Woolgar's view, the categories that Blumer talks about, categories like alcoholism and prostitution, are ontologically subjective. His complaint is that, in describing them, Blumer uses language that suggests that they are ontologically objective.

19. See, for example, Ibarra and Kitsuse (1993, 25–58). They write (p. 28): "The theory directs attention to the claims-making process, accepting as given and beginning with the participants' descriptions of the putative conditions and their assertions about their problematic character (i.e., definitions)." Ibarra and Kitsuse share with Blumer the view that sociological studies of social problems should be value free. They write (p. 27): "Sociologists [have a] right to their evaluations about those they study as well as a right to consider this or that morally offensive. But we fail to see the theoretical rationale for employing of embedding such judgments in our analytical renditions of the members' perspective."

20. The sociologist, Merton seems to think, can tell whether his subject's norms are consistent but not whether any one of them is reasonable; he can tell whether there is a gap between their heavy drinking and their norms of moderation but not whether those norms are desirable; he can explain why they disapprove of homosexuality but not whether their disapproval is warranted.

21. For a discussion of different conceptions of rationality, see Schmidtz (1993) and Anderson (1993).

22. See, for example, Anderson (1993, 104–11).

REFERENCES

Alvarez, R. 2001. "The Social Problem as an Enterprise: Values as a Defining Factor." *Social Problems*, 48, pp. 3–10.

Anderson, E. 1993. *Values in Ethics and Economics*. Cambridge, MA: Harvard University Press.

Becker, H. 1963. *Outsiders: Studies in the Sociology of Deviance*. New York: Free Press.

Best, J., ed. 1995. *Images and Issues: Typifying Contemporary Problems*. New York: Aldine de Gruyter.

Blumer, H. 1971. "Social Problems as Collective Behavior." *Social Problems*, 18, pp. 298–306.

Bowles, G., and Renate Kline, eds. 1983. *Theories of Women's Studies*. London: Routledge and Kegan Paul.

Covert, C. 2000. "Thoroughly Modern 'Eve,'" *Minneapolis Star Tribune*, December 1, 2000, p. E29.

Davis, K. 1966. "Sexual Behavior," in Merton and Nisbet (1966).

Gouldner, A. W. 1962. "The Myth of a Value-Free Sociology." *Social Problems*, 24, pp. 199–213.

Gouldner, A. W., and S. M. Miller. 1965. *Applied Sociology: Opportunities and Problems*. New York: Free Press

Gross, L. 1965. "Values and Theory of Social Problems," in Gouldner and Miller (1965).

Holstein, J., and Gale Miller, eds. 1993. *Reconsidering Social Constructionism*. New York: Aldine de Gruyter.

Ibarra, P. R., and J. I. Kitsuse. 1993. "Vernacular Constituents of Moral Discourse: An Interactionist Proposal for the Study of Social Problems," in Holstein and Miller (1993).

Lemert, E. M. 1951. *Social Pathology*. New York: McGraw-Hill.

Lewin, T. 2000. "Cut Down on Out-of-Wedlock Births, Win Cash," *New York Times*, September 24, 2000, section 4, p. 5.

Lowney, K. S., and J. Best. 1995. "Stalking Strangers and Lovers: Changing Media Typifications of a New Crime Problem," in Best (1995).

Merton, R. K. 1966. "Social Problems and Sociological Theory," in Merton and Nisbet (1961).

Merton, Robert, and Robert A. Nisbet. 1961. *Contemporary Social Problems*. New York: Harcourt, Brace and World.

Mies, M. 1983. "Towards a Methodology for Feminist Research," in Bowles and Klein (1983). Pfohl, S. J. 1977. "The 'Discovery' of Child Abuse." *Social Problems*, 24, pp. 310–23.

Proctor, R. 1991. *Value-Free Science? Purity and Power in Modern Knowledge*. Cambridge, MA: Harvard University Press.

Root, M. 1993. *Philosophy of Social Science: The Methods, Ideals, and Politics of Social Inquiry*. Oxford: Blackwell.

Root, M. 2000. "How We Divide the World." *Philosophy of Science*, 67, Proceedings, S628–39.

Rose, A. M. 1971. "History and Sociology of the Study of Social Problems," in Smigel (1971).

Schmidtz, D. 1993. *Rational Choice and Moral Agency*. Princeton, NJ: Princeton University Press.

Searle, J. R. 1995. *The Construction of Social Reality*. New York: Free Press.

Smigel, O., ed. 1971. *Handbook on the Study of Social Problems*. New York: Rand McNally.

Spector, M., and J. I. Kitsuse. 1987. *Constructing Social Problems*. New York: Aldine de Gruyter.

Tallman, I., and R. McGee. 1971. "Definition of a Social Problem," in Smigel (1971).

Tannenbaum, F. 1951. *Crime and Punishment*. New York: McGraw-Hill.

Thomas, W. I., and F. Znaniecki. 1927. *The Polish Peasant in Europe and America*. New York: Knopf.

Weber, M. 1968. *The Methodology of the Social Sciences*, tr. and ed. Edward Shils and Henry Finch. New York: Free Press.

Woolgar, S., and D. Pawluch. 1985. "Ontological Gerrymandering: The Anatomy of Social Problems Explanations." *Social Problems*, 32, pp. 214–27.

THREE

COMING TO TERMS WITH THE
VALUES OF SCIENCE: INSIGHTS
FROM FEMINIST SCIENCE
STUDIES SCHOLARSHIP

Alison Wylie and Lynn Hankinson Nelson

3.1 Nozick's Not-So-Neglected Third Option

It was clear that a sea change was under way in philosophical thinking about values and science when Robert Nozick delivered his presidential address to the American Philosophical Association in 1997.[1] How should we reconceptualize objectivity, he asked, given the challenges posed by a growing body of research—historical, sociological, and philosophical— that demonstrates just how deeply and pervasively the sciences are infused by contextual values?[2] Nozick was uncompelled by the two lines of response he found typical of philosophers. For the most part, he observed, philosophers are inclined to adopt a defensive stance, blocking any implication that local failures compromise objectivist ideals; they insist that science is objective *despite* the influence of intrusive values. At the same time, and in reaction against this response, some few have embraced, with a certain relish, the conclusion that science cannot be considered objective in any meaningful way. Nozick found both alternatives wanting and made the case for a neglected third option: that science is objective *because* of the values with which it is infused.

As a description of the polarized debates that arise whenever the subject of values in science is broached, Nozick's account is unexceptional.

But he obscures more than he illuminates when he claims to have identified a new way forward. The neglected third option that he purports to rescue from obscurity has been actively explored in a wide range of philosophical contexts. Bernstein had argued, fifteen years earlier, that options "beyond objectivism and relativism" were clearly immanent in the trajectories of philosophical traditions as different as history and philosophy of science, philosophy of social science, and philosophical hermeneutics (1983). Within philosophy of science, the family of research programs that have flourished under the rubric of social naturalism share a broad commitment to sidestep the dilemmas generated by abstract ideals of value neutrality and the skeptical challenges they attract; the goal is to understand, in empirical detail, how science is actually practiced, how its successes and failures are actually realized.[3]

Among the (social) naturalizers who have most actively explored Nozick's third option are feminist theorists and critics of science. This may seem counterintuitive. In exposing widespread, largely unremarked androcentric and sexist bias, feminists demonstrate that good science, even our best science, can be deeply structured by the values and interests of its makers; scientific practice, even at its most rigorous, is not always or automatically self-correcting (see, e.g., Harding 1986, 19, 102–5; Fausto-Sterling 1985, 7–9; Wylie 1997a). But by no means does this entail the corrosive relativism so often attributed to feminist and other science critics. Feminists engage the sciences not only as critics of bias and partiality but also as practitioners who recognize that systematic empirical inquiry has an indispensable role to play in understanding and changing oppressive conditions. They rely on the tools of scientific inquiry to expose biases that arise, directly and indirectly, from unexamined assumptions about sex/gender systems, and they make effective use of these tools to explore new lines of inquiry that open up when they bring a feminist angle of vision to bear on their various fields. Contra critics who impugn these thriving research programs as "just political," the epistemic stakes are high precisely because of the political commitments that animate them; the goal of understanding and changing conditions of life that disadvantage women requires as much empirical accuracy and explanatory precision as scientific inquiry can afford. It is not surprising, then, that, however skeptical feminists have been about specific claims to objectivity and value neutrality, rarely do they endorse a wholesale rejection of the constitutive values—the epistemic standards and ideals—that are typically assumed to characterize and distinguish scientific practice.

The question then arises: What are the research-shaping commitments that feminists bring to the sciences (specifically the social and life sciences, for purposes of this discussion), and how do these articulate with conventional epistemic values that are presumed to be constitutive of the scientific enterprise? Although they vary a great deal from field to field and depending on the specific form of feminism embraced by the practitioners in question, it is possible to identify a loosely articulated set of empirical, ethical, and political commitments that comprise what Longino refers to as feminist community values. These are characterized, minimally, by two metaprinciples: a "bottom line maxim," as Longino describes it, that captures a commitment to "prevent . . . gender from being disappeared," and a principle of epistemic provisionality, according to which any more specific feminist guidelines and assumptions must be held open to revision (Longino 1994, 481; Longino 1987; Nelson 1990, 29–37; Wylie 1994, 1997a). In a great many contexts, the partisan concern to explore the difference that sex/gender systems make to our lives and to our scientific understanding has put feminists in a position to make substantial contributions to the research fields in which they work. It has been the catalyst not only for critical and corrective interventions but also for strikingly original lines of inquiry that had not previously been considered and that often reshape disciplinary practice well beyond the specific research programs in which feminists are directly engaged.[4] These very successes reinforce Nozick's central insight: that contextual values are not only or always compromising of scientific practice but can have a positive impact as well. A model of the relationship between science and values that requires strict value neutrality, the elimination of contextual values from well-functioning (objective) science is clearly inadequate.

From the outset, feminists have been alert to the philosophical tensions generated by their jointly critical and constructive engagement with the sciences. In reflection on the epistemic implications of their practice, they invariably grapple with the specter of corrosive relativism, but even self-identified postmodern theorists such as Haraway (1991) insist that this is as untenable as naive objectivism. The challenge has been to understand, in conceptually nuanced and empirically specific terms, the difference that contextual values make to science, both positive and negative (Nelson and Wylie 2004). Two key questions are important in this context:

1. How are we to reconceptualize objectivity so as to undercut polarizing, dichotomous patterns of argument that assume that any recognition of the play of contextual values in science is inimical to ideals of objectivity?
2. How exactly do values of various kinds make a difference to objectivity, positive and negative?

We focus here on the second question, considering two case studies that illustrate a number of kinds of constructive difference that (feminist) contextual values can make to scientific practice. But we take as our point of departure a prospective answer to the first question, an analysis of objectivity proposed by Lloyd in which distinctions are drawn between several different kinds of referent to which objectivity can be attributed or denied (Lloyd 1995; see also discussion in Wylie 2003). Lloyd is chiefly concerned with objectivity as an attribute of the objects of knowledge. This is objectivity in a quasi-metaphysical sense, attributed to entities and events that are presumed to exist, and to have the distinctive properties they have, independent of any observer: entities and events that have "objective reality," the "really real," as Lloyd puts it. We focus on tensions that arise between two further senses of objectivity Lloyd discussed: objectivity as an attribute of knowers and objectivity as an attribute of the knowledge they produce. Objectivity in the second sense is often equated with a requirement of value neutrality or disengagement in epistemic agents, which is presumed, in turn, to be indicative of (or a necessary condition for) the objectivity of the knowledge claims they produce or endorse. When objectivity is attributed to knowledge claims, it is an honorific that picks out those characterized by a loosely defined family of epistemic virtues. Standard lists of such credibility-conferring virtues include empirical adequacy in at least two senses—fidelity to particulars, as opposed to breadth of applicability (a capacity to "travel" as Haraway describes it [1991])—internal coherence, external consistency with other established bodies of knowledge, and explanatory power under extension to a specified range of cases or domains, in addition to other more heuristic and aesthetic virtues, such as simplicity, elegance, and manipulability (Longino 1995; Kuhn 1977).

When these latter two senses of objectivity are distinguished, it becomes clear that the objectivity (qua neutrality) of an epistemic agent by no means guarantees the objectivity (qua epistemic credibility) of knowledge claims. In the first set of cases we consider, from archeology, the value commitments of knowers—their nonobjectivity in the second sense—is instrumental in improving objectivity in the third, epistemic sense. It illustrates the central points made previously about feminist contributions to the sciences, with particular attention to the ways in which a sensitivity to gender-conventional assumptions can enrich a research program empirically. When objectivity is considered as an attribute of knowledge claims, the heterogeneous epistemic virtues in terms of which it is assessed are unavoidably ambiguous in their implications for practice (they require interpretation), and they are often incapable of simultaneous satisfaction (they must be weighed against one another). Kuhn took this to be one of the important implications of the historical cases analyzed in *The Structure of Scientific Revolutions*;[5] there are always trade-offs to be made between the epistemic attributes of a knowledge claim that establish its credibility. Feminist philosophers of science extend this insight, arguing that contextual values necessarily enter into the judgments researchers make about what counts as adherence to these values and how to set priorities when they come into conflict (e.g., Longino 1995 and 1997). Our second case study, from biology, illustrates the play of contextual factors in this process of interpreting and weighing constitutive epistemic values; here a feminist angle of vision generates a fruitful reassessment of accepted metamethodological conventions that have had the (unintended) consequence of allowing sex/gender stereotypes to persist in the explanatory framework of the field.

3.2 How Contextual Values Can Promote Epistemic Values

3.2.1 *First Case Study: How Feminist Standpoints Have Enriched Archaeological Practice*

Recent developments in American archaeology offer a number of examples in which contextual values play an instrumental role in improving the quality of scientific research, judged by quite conventional epistemic standards.[6] In *Toward a Social History of Archaeology in the United States* (1995), Patterson documents the difference the G.I. Bill made to the class structure of the discipline and the ways this has, in turn, changed the questions asked, the hypotheses considered, the skills valued, and the relative influence of distinct regional traditions of archaeological practice in the United States. This legacy is by no means an unmixed blessing, but Patterson makes a convincing case that it has played a crucial role in disrupting elitist assumptions and significantly broadening the scope of North American archaeology.

Another major demographic change in the recent history of North American archaeology has been the influx of women into the field in the 1970s and 1980s. In the space of a decade, the representation of women in the field nearly doubled,[7] and by the end of the 1980s, when the first of these cohorts came of professional age, the first substantial work on questions about women and gender began to appear (Nelson, Nelson, and Wylie 1994). These developments appear much later in archaeology than in neighboring fields (like sociocultural anthropology and history), where vigorous traditions of feminist research had taken shape since the turn of the 1970s. The first paper explicitly advocating feminist approaches to archaeology appeared in 1984 (Conkey and Spector), but it wasn't until 1988–1989 that the first exploratory conferences on "the archaeology of gender" were organized, the proceedings of which began to appear in the early 1990s (Gero and Conkey 1991; Walde and Willow 1991). Since that time, archaeologists have taken up a range of gender research projects, both critical and constructive, in virtually all major cultural, temporal, and geographical areas of interest and in connection with the full range of research problems typical of the field (see summaries and overviews in Wright 1996; Nelson 1997; Claassen and Joyce 1997; Hays-Gilpin and Whitely 1998; Gilchrist 1999).

From the outset, internal commentators have registered disappointment that archaeologists working in this area are often reticent to identify themselves or their research as feminist or to engage established traditions of feminist research in related fields (Hanen and Kelley 1992), a pattern that has persisted (Conkey and Gero 1997; Engelstad 2004). Nonetheless, a case can be made that the growing body of archaeological research in the "gender genre" does embody Longino's "bottom line" maxim for feminist research: the commitment to "mak[e] gender a relevant axis of investigation" (Longino 1994, 481; Wylie 2001). The content of the work that has appeared, as well as the timing and circumstances that gave rise to it, suggest that, although explicitly feminist influences are muted, a standpoint of gender sensitivity has played a crucial role in shaping these developments (Wylie 1997b; Hanen and Kelley 1992). Those drawn to archaeological research on gender were, in the first instance, predominantly women who had entered archaeology at a time when their rapidly increasing representation disrupted the gender status quo of the field; they are often quite explicit in rejecting the gender norms and stereotypes that continue to circumscribe their participation in the profession, even as they distance themselves from feminist scholarship and activism. As diffuse as it is, this heightened awareness of the gender norms and assumptions that structure everyday life—this gender sensitivity—has inclined a growing number of women (and some men) in this cohort to identify and question gender-conventional assumptions that underpin archaeological thinking about the cultural past. This has reframed and enriched archaeological research at all levels of practice, from the most empirical through to the most interpretive and theoretical.

Consider, first, the impact that a standpoint of gender sensitivity has had on the evidential basis of the field. Often the commitment not to "disappear gender" is interpreted in quite literal terms; it takes the form of a newly insistent interest in documenting the presence, contributions, and activities of women, which, in turn, directs attention to dimensions of the archaeological record that have been overlooked or interpreted in terms of normatively ethnocentric and androcentric assumptions about gender roles and relations. Although internal critics object to the limitations of such an approach, it has had significant impact, reshaping not just low-level reconstructive claims but also, through them, more ambitious explanatory models and framework assumptions. In some cases, the challenge has been to counteract the assumption that women's activities are archaeologically inaccessible because they are associated with ephemeral and perishable materials. In fact, as the growing corpus of work on netting and basketry demonstrates, there is much potential here that has been underexploited. Systematic examination of surviving fragments of these materials reveals a sophistication and complexity in traditions of manufacture that had not been recognized even when the primary data had been collected (e.g., Bernick 1998), and these insights are further enhanced by indirect lines of evidence that are now being recovered, for example, through the analysis of bone awls that would have been used to make nets (Dobres 1995; Soffer 2004), the imprint of twine, and images of woven clothing and hair nets that survive in other media (Soffer, Adovasio, and Hyland 2001). Similar strategies have significantly expanded the database relevant for reconstructing female-associated subsistence activities, for example, through the analyses of breakage patterns in bones that can be attributed to secondary food processing rather than primary butchering (Gifford-Gonzalez 1993) and in renewed attention to edgewear patterns in utilitarian stone tools that reveal evidence of foraging and resource exploitation activities presumed to be women's domain, activities that had largely been ignored in preference for analysis of male-associated hunting tool assemblages (Gero 1991, 1993).

In a parallel shift of emphasis, the study of skeletal remains has been fruitfully reoriented to the analysis of dietary, disease, and activity profiles that hold the promise of bearing witness to archaeologically enigmatic dimensions of the cultural past: social schisms and assimilations, enslavement, and differentiation along lines of status and factional difference. Researchers working from a standpoint of gender sensitivity have focused attention on how dietary status and vulnerability to violence differ along sex/gender lines and on the possibility that the differential development of muscle attachments and evidence of stress may reflect sex-specific patterns of repetitive activity (see the summary provided by Cohen and Bennett 1993 and also Bentley 1996). Sometimes this requires not so much new lines of analyses that produce data not previously considered as the critical reinterpretation of data already in hand. One relatively straightforward example comes from a reanalysis of skeletal material recovered from archaeological contexts in Australia that showed a vast preponderance of male specimens (Donlon 1993). Donlon found implausible the standard explanations for this imbalance—demographic anomalies, differential preservation, curatorial practices—and argues that these skewed sex ratios most plausibly reflect systematic errors in sex identification due to reliance on measures of "robustness" that presuppose ethnocentric stereotypes of physical dimorphism. Given what is known historically and ethnographically about the activities typical of Aboriginal women, Donlon observes, it is to be expected that they would show much the same levels of skeletal robustness as their male counterparts; it is a mistake to project onto prehistoric Aboriginal foragers the norms of gender-segregated physical activity that are conventional in sedentary, largely urban, middle-class contexts.

As this last case suggests, the potential for enriching the evidential base in ways that will counteract the disappearing of women and gender often depends on a reassessment of the background knowledge—the auxiliary hypotheses or linking principles ("middle range theory," as archaeologists refer to it)—on which archaeologists depend to interpret their data as evidence. A gender-sensitive standpoint has been the catalyst for a number of research programs in experimental and ethnoarchaeology designed to reassess the background assumptions that underpin androcentric or sexist interpretations and to provide the resources necessary for recognizing evidence of gender roles and dynamics not previously investigated. Perhaps the best known are feminist critiques of "man the hunter" models of prehistoric subsistence practice that depend, in part, on a systematic reassessment of the ethnographic sources that were the basis for interpretations of the archaeological record, as well as the associated evolutionary scenarios. Although the hunting practices typical of men had long been the focus of ethnographic attention and the basis for generalized descriptions of these societies, a new generation of field studies in the 1960s and 1970s, some of them explicitly feminist in orientation, established that in many settings the bulk of the dietary intake came from small game and plant material provided by the "gathering" activities of women (see, e.g., the summary given by Slocum 1975). Although this shift in understanding of the background (analogical) sources very effectively challenged conventional (androcentric and sexist) assumptions about prehistoric sexual divisions of labor and the models of human evolution based on them, they leave unanswered a range of more narrowly specified questions about the material correlates of specific gender roles and divisions of labor among foragers—questions that are crucial for developing more credible archaeological models. To address these, a thriving tradition of ethnoarchaeological research has taken shape in the last thirty years that is providing a much enriched understanding of the social and organizational complexity of foraging practice, some of which is informed by a standpoint of gender sensitivity. For example, in a study of the hunting practices of the subarctic Dene, Brumbach and Jarvenpa (1997) focus on the understudied role of women; they document much greater diversity in the gender composition of hunting parties than had been recognized and demonstrate that women play a prominent role in some forms of hunting. In addition to its implications for understanding historic shifts in subsistence practice among the Dene, these results also suggest, more broadly, that archaeologists should reassess standard ascriptions of function to sites and artifacts where these depend on an identification of hunting activities with men, sharply segregated from the domestic activities associated with women.

These are, then, cases in which a standpoint of gender sensitivity— a commitment to ensure that gender (and women) are not disappeared— has provoked a reexamination of disciplinary conventions about what can or should be studied archaeologically. This, in turn, directs attention to new ranges of data and new possibilities for interpreting (or reinterpreting) archaeological data that shift the evidential horizons of the discipline as a whole. Sometimes the result is a reassessment of androcentric models that inverts gender conventions, so that women are recognized to have played a central role in domains of cultural life, and in processes of cultural change, that had typically been attributed to men. One such example is Sassaman's argument that the transition from mobile to sedentary prehistoric societies in North America should not be attributed automatically to changes in the organization of hunting practices associated with men: "considering . . . that the change coincides with the adoption of pottery, technology usually attributed to women, an alternative explanation must be considered" (1992, 249). In other cases, the effect of bringing women and gender into focus archaeologically is to undermine gender categories more fundamentally, suggesting that it may be a mistake to assume any gender segregation of activities along lines familiar from contemporary contexts. This turn was taken in later rounds of the debate about theories of human evolution when feminist critics of the 1990s questioned the wisdom of countering "man the hunter" with "woman the gather" scenarios; these, they argued, simply invert naturalized (ethnocentric) gender categories that ethnographic and primatological research suggests should be systematically reassessed (Sperling 1991). In an archaeological context, McGuire and Hildebrandt argue for a reinterpretation of California sites that provide evidence of long-term reliance on and change in the practice of milling acorns and other plant resources. They argue that this evidence of milling should be recognized to reflect "a consistent pattern of relatively undifferentiated gender roles," in which case the shift from millingstone-handstone technology to bedrock mortars, and associated changes in social organization and mobility, should not be attributed exclusively to women (1994, 51): "if we have learned anything [from] the emerging feminist critique of modern archaeology, it is perhaps the danger in viewing gender relationships as static, or at least limited in range" (1994, 52).

At a further remove from the interpretation of specific types of site or classes of archaeological material, gender-critical perspectives have suggested some highly productive lines of comparative analysis, juxtaposing data sets that had typically been treated in isolation.[8] In one widely discussed example, Hastorf (1991) notes that while skeletal series excavated in the Montaro Valley had been sexed and lifetime dietary intake estimated on the basis of bone marrow (isotope) analysis, no one had noticed that male and female skeletons show a striking divergence in dietary profiles (specifically, their levels of maize consumption) through the period when the Inka first appeared in the Montaro Valley (Hastorf 1991). In a similar vein, Brumfiel (1991) identifies subtle but significant shifts in the distribution of well-documented artifacts related to food and cloth production between urban and hinterland sites in the central Mexico Valley through the period when the Aztec state was consolidating its system of tribute payments in the region. The composition of assemblages typical of sites closest to the urban centers suggests that women's household labor was increasingly invested in the production of labor-intensive, portable food (tortillas and other foods based on ground corn), supplying an urban labor force that sold its services in exchange for the required tribute cloth. By contrast, the domestic labor of women in outlying sites was reallocated in the opposite direction; the artifacts and features associated with the production of highly processed foods declined in favor of stew-pot preparations, while evidence for the production of cloth intensified.

Both Hastorf and Brumfiel use this comparative reanalysis of existing data sets to make a case for rethinking dominant explanatory models of the development and internal dynamics of prehistoric Inka and Aztec states. They argue, in rather different (and context-specific) ways, that our understanding of state formation processes is seriously flawed if these are identified exclusively with structural dynamics operating in the public, political sphere associated predominantly with male elites. It cannot be assumed, they insist, that household structure and the organization of domestic and reproductive labor are a stable (quasi-natural) social substrate, unaffected by and inconsequential to the business of state building. Hastorf combines the evidence of sex-divergent dietary profiles with analyses of paleobotanical remains and food-processing features from a sequence of house floors, which suggest that household organization and the focus of domestic labor changed substantially to meet the demands of the Inka as they extended the influence of their empire into the highland Andes. Likewise, Brumfiel argues that the tribute system of the Aztec state depended fundamentally on a redeployment of domestic labor that changed gender roles and power dynamics at the household level. In neither case can it be assumed that sex/gender systems are a (biological) foundation that remains stable as the fortunes of states rise and fall.

In each of these cases, it is the use of well-established databases, techniques of analysis, and inference strategies that lends credibility to arguments for reassessing the gender-normative assumptions that have informed archaeological research in a very wide range of contexts, at every level of practice; these are evidentially grounded challenges to research traditions that have left women and gender out of the account or interpreted them in ethnocentric terms. As such, they improve archaeological understanding in quite conventional terms, on dimensions captured by several of the epistemic virtues characteristic of objectivity (in the third sense). They enhance empirical adequacy by expanding the range of data and enriching the interpretive resources that inform local (event and context-specific) archaeological reconstruction, and they use these resources to redress a gendered pattern of gaps and distortions in reconstructive models that are consequential for broader explanatory theories. But as in many other cases, these improvements cannot be attributed to a commitment to objectivity in the second sense: to freedom from the influence of contextual values. The methodological routines of mainstream archaeology—"science as usual"; good science—had systematically reproduced the lacunae addressed in these cases and showed no evidence of redirecting attention to those dimensions of the archaeological record or interpretive resources that would be instrumental in addressing them. It was the gendered interests and perspectives of a subset of practitioners, overwhelmingly women from recent professional cohorts, that provided the conceptual resources and the catalyst for these interventions; this standpoint of sensitivity to gender assumptions put them in a position to identify critical limitations and investigative possibilities that had not been recognized by colleagues for whom androcentric and gender-stereotypic assumptions were unproblematic presuppositions of the research enterprise. In this the late but vigorous development of research on women and gender in archaeology illustrates Nozick's third option: Here the intrusion of nonepistemic values and interests is by no means epistemically compromising but rather plays an epistemically enabling role.

3.2.2 *Second Case Study: Epistemic Evaluative Judgments and the Role of Contextual Values*

In our second case, we consider how feminist critiques of quite fundamental assumptions guiding research in developmental biology transformed thinking about what was, through the early 1980s, the textbook account of human embryonic sexual differentiation. In a number of respects, the standard account conformed to a broad body of research into fetal development. But explicitly feminist perspectives threw into relief widely held contextual values that shaped and organized the research on which the account drew, including the relative weight that had been accorded various epistemic values by the scientists involved.

According to the standard account, as illustrated by Tuchmann-Duplessis, Aurous, and Haegel's *Illustrated Human Embryology* (1975), sex or, more precisely, genetic sex is determined at conception: Eggs fertilized by X-chromosome-bearing sperm develop into females, and eggs fertilized by Y-chromosome-bearing sperm develop into males. In the six weeks following conception, XX and XY embryos are anatomically identical. Both develop an "indifferent" gonad that includes a set of female (Müllerian) ducts (which in the female will later develop into the uterus, cervix, oviducts, and upper vagina) and a set of male (Wolffian) ducts (which in the male will develop into vas deferens, epididymis, and ejaculatory ducts). Anatomical differentiation, the development of gonadic sex, begins in the sixth week. On this account, the genetic information present on the Y chromosome promotes the synthesis of the H-Y antigen in XY embryos; this protein, in turn, plays a role in the organization of the indifferent gonad into an embryonic testis. In addition to containing sperm-producing tubules, the embryonic testis is able to synthesize hormones. Two of these, testosterone and Müllerian inhibiting substance (MIS), function to promote additional development in a male direction. Fetally synthesized testosterone promotes growth and development of the male duct system; MIS causes the female duct system to "regress" or "degenerate."

Through the eighth week, the external genitalia of XX and XY embryos are anatomically identical. Sexual differentiation occurs in the ninth through twelfth weeks, when the bipotential genital tubercle develops into a penis or a clitoris, and tissues called "labioscrotal swellings" develop into a scrotum or large lips of a vagina. Fetal hormones were recognized to be instrumental in the sexual differentiation of the bipotential genital tubercle. According to the standard account, the fetal testis secretes dihydrotestosterone, and this influences development of the genital tubercle in a male direction. By twelve weeks, as the conclusion of the account puts things, the structures of male external genitalia are evident in XY embryos, and the structures of female external genitalia are evident in XX embryos.

This account of human sexual differentiation drew on and was consistent with a broad body of research in biology. It depended on what was known about the anatomical stages of fetal development and sexual differentiation, as well as on baseline commitments in evolutionary biology, developmental biology, and endocrinology to sexual dimorphism and the definition of sex in terms of gamete production (males are organisms that produce sperm; females are organisms that produce eggs). The emphasis on a causal role for androgens was consistent with investigations in reproductive and neuro endocrinology into the roles of prenatal androgens in brain organization in males of various species (or, as it was described at the time, how "male brains" are created). And, in general terms, its notions of gender dimorphism—specifically, that various factors coded "male" "actively" intervene in what would otherwise be a "normal" and "female" line of development—were consistent with schemas and hypotheses in developmental biology, endocrinology, and empirical psychology. So external consistency contributed to the credibility of the standard account of human embryonic sexual differentiation.

In addition, the various stages of development in human sexual differentiation had parallels in other mammalian species, and the explanation of these stages was in agreement with a broad range of observational and experimental data. In this sense, the account can be said to enjoy generality of scope and, in several ways we have noted, to display accuracy under extension to a range of cases and domains (Nozick) or to have the capacity to "travel" (Haraway). This generality of scope *in these senses* was a function of the *simplicity* of the hypothesis that the presence or absence of the Y chromosome determines the trajectory and details of sexual differentiation and development.[9] This hypothesis effectively streamlined a number of complex processes and variables. Two unidirectional courses of development are posited that are, in turn, taken to be solely determined by the presence or absence of relatively few variables (Y-chromosome-bearing sperm, the H-Y antigen, the embryonic testis, and dihydrotestosterone). In so doing, this framework streamlines explanation: Of the hormones synthesized by both XX and XY embryos — which include androgens, estrogens, and progesterone — only the first group are invoked; a male course of development occurs when these are present, and a female course in their absence. Causal relationships posited between hormones and stages of development are assumed to be linear or unidirectional (cf. Longino and Doell 1983; Longino 1990). Finally, within this framework, complicating factors in the maternal environment (e.g., the hormones derived from the placenta and maternal circulation) can be assumed to play no causal role in sexual differentiation, although scientists involved in this research expressed some puzzlement that XY embryos are not "feminized" by the abundance of estrogen and progesterone in their environment (Fausto-Sterling 1985).

For all these reasons, the standard account was widely influential and tenacious, and it generated fruitful research: of how the H-Y antigen functions to promote the development of the embryological testis; of which genes on the Y chromosome promote the synthesis of the H-Y antigen and how; and of how dihydrotestosterone promotes the growth of male external genitalia.

By the early 1980s, however, feminist biologists had begun to criticize this account, as well as the general model of fetal development that shares its central presumptions. They drew attention to the consequences, for other epistemic values, of putting particular weight on the epistemic values associated with the capacity of a knowledge claim "to travel," noting the role of contextual values in such trade-offs. We take feminist critics of the standard account to make two arguments. One is that the value researchers had attributed to simplicity and to generality of scope *in one sense* — applicability across species — brought too high a cost in terms of generality of scope *in another sense*, namely, a lack of explanation for *female* fetal development and for the role of hormones and organs coded female, as well as the X chromosome, and this selective emphasis on scope, in turn, brought too high a cost in terms of explanatory power and empirical adequacy. The second argument is that contextual values, albeit unrecognized by the scientists involved, were a factor in these choices. We will argue, further, that it was the contextual values that feminist biologists brought to bear that *revealed* these trade-offs among epistemic values and also revealed the role of (submerged) contextual values.

The trade-offs between epistemic virtues that we are positing are perhaps most obvious when we consider the argument made by feminist biologists that the textbook account of sexual differentiation is an account of *male* sexual differentiation, not *embryonic* sexual differentiation. By the early 1980s, feminists in biology had begun to analyze the role and consequences of what John Money referred to as "the Adam principle," a commitment he attributed to fetal development research, according to which something must be *added* to an embryo to make it a male (Money 1970, 5). The corollary of this hypothesis, feminists noted, is that the female and female fetal development are, respectively, the default state and default trajectory of fetal development (Fausto-Sterling 1985, Gilbert et al. 1988).

Neither "the Adam principle" nor its corollary rules out the need to investigate female development. But as feminist biologists recognized, the "presence" and "absence" metaphors informing the textbook account are aptly ascribed to the account itself. Largely absent are discussions of or investigation into the role of entities classified as "female": the hormones estrogens and progesterone, the fetal ovary, the maternal environment, and to some extent, the X chromosome. Until "a positive role for estrogen began to creep into parts of the literature," research on the role of estrogen focused largely on the question of why the developing male embryo was not feminized by it (Fausto-Sterling 1987, 67). Mentions of the possible role of hormones secreted by fetal ovaries, as occurs in a review article quoted later, were anomalies and also indicated the lack of information (and lack of relevant research) concerning such a role. "Embryogenesis normally takes place in a sea of hormones . . . derived from the placenta, the maternal circulation, the fetal adrenal glands, the fetal testis, and *possibly* from the fetal ovary itself . . ." (emphasis added; Wilson, George, and Griffin, "The Hormonal Control of Sexual Development," quoted in Fausto-Sterling 1985, 81).

Further, according to the standard account, given the absence of so-called male factors, the indifferent gonad of the XX embryo "develops into the internal female reproductive system," the male ducts "degenerate," and the female structures of external genitalia "become *evident* by twelve weeks."[10] As Fausto-Sterling noted, "The view that females develop from mammalian embryos deficient in male hormone seems, oddly enough, to have satisfied the scientific curiosity of embryologists for some time, for it is the only account one finds in even the most up-to-date texts. . . . How does [female development] happen? What are the mechanisms?" (Fausto-Sterling 1987, 66). Indeed, she argues, what resulted "in a supposedly general account of the development of the sexes . . . [was] in actuality only an account of male development" (Fausto Sterling 1987, 64). For at about the same time that the XY gonad begins to make testosterone, the XX gonad "appears to begin synthesizing large quantities of estrogen." "Just what," Fausto-Sterling asked, "does all that estrogen do?" (Fausto-Sterling 1985, 81).

So far, it is clear that feminist biologists were challenging the empirical adequacy, explanatory power, and generality of scope of the textbook account of sexual differentiation because it failed to explain female sexual differentiation or, at the very least, failed to investigate the possible role of entities coded "female" in female development. But feminist biologists further argued that lack of information about the mechanisms of female sexual differentiation, about the role of hormones coded "female," and about the role of the maternal environment also raises questions about the empirical adequacy of the textbook account of male sexual differentiation.

Consider, for example, the role of the X chromosome in embryonic sexual differentiation. It seems clear that the Y chromosome promotes the synthesis of the H-Y antigen that, in turn, promotes the organization of embryonic testes. The testes, in turn, are able to synthesize testosterone and MIS, and so forth. But it is *not* the Y chromosome that enables the synthesizing of testosterone and MIS. It is genetic information on the X chromosome and *on one or more* of the twenty-three pairs of nonsex chromosomes that code for androgens and estrogens. So while it was clear that the Y chromosome is somehow involved in the selective translation of some of this information, in focusing exclusively on this component of the process, researchers learned little about the mechanisms that determine "identical road maps for sexual development" in XX and XY embryos (Fausto-Sterling 1985, 79–85).

Feminist biologists also argued that conclusions cannot be drawn from research that apparently establishes the effects of prenatal androgens until a similar amount of research is done on the organizing effects of prenatal estrogens. The fetal environment is rich in both, males and females synthesize both (it is the amounts that differ), and there are continuous conversions among the three families of sex hormones (e.g., Bleier 1984, 1988; Fausto-Sterling 1985). The difficulties in isolating the effects of these hormones are attested to by the recent reversal of claims that an "organizing effect" on fetal rat brains that researchers had attributed to androgens and linked to behavior they characterized as "masculine" is now attributed to estrogen converted from testosterone by brain cells (Fausto-Sterling 1987; Nelson 1996). And, as Fausto-Sterling points out, the "sea of hormones" in which embryogenesis occurs includes hormones derived from the placenta and the maternal circulation, but relatively little attention has been paid to the possible role of these hormones in either male or female development.[11]

Finally, feminist biologists criticized the unidirectional model of male sexual differentiation and cited experimental results that indicate complex and often nonlinear interactions between cells and between cells and the maternal and external environments during every stage of fetal development (Bleier 1988; Fausto-Sterling 1985; Hubbard 1982).

It is on the basis of these several lines of argument that feminist biologists challenged the empirical adequacy and explanatory power of the textbook account of sexual differentiation (Bleier 1984; Fausto-Sterling 1985, 1987; Longino and Doell 1983). In the process, they directed attention to a number of underlying evaluative judgments in which simplicity is preserved at the cost of explaining the process of female sexual development and thus at the cost of generality of scope, empirical adequacy, and explanatory power.

Perhaps less obvious, because it is so deeply embedded in the biological sciences and conventional commonsense thinking, is how the commitment to sexual dimorphism serves to simplify what might otherwise be recognized as more complex, and also relevant, factors and processes. On the one hand, the commitment is consistent with the definition of the sexes as "egg-producing" and "sperm-producing" organisms, which has guided fruitful research. Using it, for example, males and females can be assumed to provide a natural baseline for investigating the relationships between the hormones, neural events, and behavior with which reproductive and neuro endocrinology are concerned (cf. Longino

1990; Nelson 1996). On the other hand, feminist biologists argue that the assumption of sexual dimorphism is responsible for oversimplification in fetal development research and the textbook account of sexual differentiation. As we have seen, it organizes the classification of the hormones synthesized during fetal development as "male" and "female" and bifurcates the investigation of their effects, despite the reasons for thinking that the bifurcation is problematic noted earlier.[12] In addition, Fausto-Sterling challenges the assumption of sexual dimorphism on the grounds that "biologically speaking, there are many gradations running from female to male; and depending on how one calls the shots, one can argue that along that spectrum lie at least five sexes—and perhaps even more" (Fausto-Sterling 1993, 21; cf. Bleier 1984). More specifically, she maintains that sexual dimorphism bifurcates into two categories organisms that actually form a continuum in terms of physiological, chromosomal, and hormonal traits, obscuring complexities and processes that might be visible with different sorting. Some individuals are classified, on this basis, as "intersexed" and are studied as "abnormalities," as opposed to being recognized as distinct sexes worthy of investigation in their own right.

We have so far emphasized ways in which feminist critics reveal underlying epistemic judgments that privilege simplicity and generality of scope (in the sense of cross-species applicability) over empirical adequacy, explanatory power, and generality of scope in another sense. We now consider how these critiques demonstrate that contextual values can be a factor in the judgments by which epistemic values are interpreted and prioritized.

One of the contextual values that feminist biologists have identified as guiding research into sexual differentiation is androcentrism, a value whose role in scientific theorizing was largely unrecognized prior to feminist science scholarship but for which there is now compelling evidence.[13] Examples of androcentrism in the present case include the emphasis on male development and relative lack of interest in the mechanisms of female development, the association of males (and entities associated with them, e.g., the Y chromosome and testosterone) with activity and females (and entities associated with them, e.g., the fetal ovary and estrogens) with passivity, and the association of males with "presence" and of females with "lack." These associations are reflected in the assumption that there is "an intrinsic tendency [of the fetus] to

develop according to a female pattern of body structure and behavior"
that male factors "must actively counteract" (Goy and McEwen 1980)[14]
and in the assumption that the fetal ovary, estrogens, and maternal environment have no causal role in sexual differentiation and development. Androcentrism is also evident in the names given to the sex hormones: *androgen*, from the Greek *andros* and Latin *generare*, means "to
make a male"; *estrogen*, from the Latin *oestrus*, means "gadfly" or
"frenzy." As Fausto-Sterling notes, "*gynogen* would be etymologically
and biologically correct as a counterpart to *androgen* . . . [but it] cannot
be found in biological accounts of sexual development (or, for that matter, in any dictionary)" (Fausto-Sterling 1987, 66). The issue Fausto-
Sterling raises is not, of course, about language. The name given to hormones coded female reflects the lack of causal role attributed to them,
as well as historical sexism.

Finally, it is clear that the contextual values brought to bear by feminist scientists—including the bottom-line maxim that gender not be
disappeared, as well as more specific injunctions against the kinds of
oversimplification that animate the interest in reductive and determin-
ist accounts of sex difference[15]—played a role in their evaluative judgment that the costs of the trade-offs among epistemic values characteristic of the standard account were unacceptable. Although other influences
might eventually have drawn attention to factors and processes that
were largely ignored in the standard account, there can be no question
that feminist values informed the critiques that were instrumental in exposing the role of (androcentric) contextual values in the evaluative
judgments that underpin the standard account and in identifying a host
of factors to which an empirically adequate account of human *embryological* development would need to attend. These values contributed
to both the critical and the constructive elements of feminist engage-
ments with research on human fetal development.

3.3 Implications for Objectivity

The cases we have considered, which we take to be representative, are
ones in which feminist, or gender-sensitive, scientists have identified
ways in which contextual values have shaped and often circumscribed
the questions pursued, the observations made, the interpretations of

empirical results, and the hypotheses generated in research programs widely regarded as well conceived, rigorous, and productive. Typically, these effects arise and persist because scientists are unaware of the values informing the research traditions in which they are educated and in which they work, particularly when these involve a gendered dimension; conventional assumptions about gender roles and relations, attributes, and identities are so deeply entrenched and so pervasive that it is not surprising that they have been so widely taken for granted. The further significance of the critiques we have summarized and others like them is that sometimes crucial advances in the sciences depend on the intrusion of contextual values; the various and contingent roles they play in science are not exclusively a matter of contamination and compromise.[16] The problem we face, if we are committed to understanding and improving scientific practice, is no longer that of cleansing science of intrusive values but, rather, that of determining what kinds of contextual factors, under what circumstances, are likely to advance the cause of science in specific ways, where the goals and standards of science are themselves evolving and open to negotiation.

Although these cases provide no comprehensive catalogue of answers to the question of exactly how values of various kinds make a difference to objectivity, they do suggest a strategy of response. They make it clear that this question cannot be pursued from philosophical armchairs; adequate answers require systematic, socially naturalized investigation of the interplay of epistemic and contextual values in specific scientific disciplines and programs of research. Such jointly philosophical, historical, and sociological-anthropological research must be open-ended methodologically and theoretically; it cannot be assumed in advance that epistemically consequential judgments are determined by evidence and good reasons alone or, for that matter, that they are ultimately a function of social conventions and power dynamics. The point of departure for such work, reinforced by the results of feminist explorations along lines we describe here, is an appreciation that it is a contingent matter what counts as adherence to objectivity-conferring epistemic virtues; how they are interpreted and weighed against one another reflects not only specific histories of research practice in which epistemic conventions are refined but also pragmatic goals and technical constraints, social conventions and political interests. It follows from this that it is also a contingent matter what range of contextual and constitutive factors—what collective or individual values, animating interests, forms of community organization—will be optimal for improving objectivity in the products of specific kinds of inquiry. The challenge for a naturalized (and socialized) philosophy of science committed to exploring Nozick's third option is, then, to delineate the effects that particular institutional structures, community dynamics, contextual interests, and values have on the forms of practice they (partially) constitute.

This naturalizing proposal suggests a way of refining answers to the first of the two questions we posed—the question of how best to reconceptualize objectivity—that are currently emerging from feminist science practice and from feminist philosophy of science. Insofar as we construe this as a question about the epistemic stance appropriate to scientists, as individuals or communities, the complex relationships between science and values revealed by feminist analysis suggest the need to value critical reflexivity rather than value neutrality. The prospects for enhancing the objectivity of scientific knowledge are most likely to be improved not by suppressing contextual values but by subjecting them to systematic scrutiny; this is a matter of making the principle of epistemic provisionality active.[17] The kinds of critical analysis embodied in feminist and other socially naturalized programs of science studies shift from background to foreground; they are properly regarded not as ancillary to the primary business of actually doing science but as a crucial constituent of good scientific practice.

NOTES

1. For analysis of the published version of this address (Nozick 1998), see Wylie (2000).

2. In this discussion, we make use of the standard distinction between contextual values and epistemic or constitutive values as drawn, for example, by Longino (1990, 4–7): the distinction between factors or considerations that are conventionally treated as appropriately external to science (also referred to as noncognitive or nonepistemic interests and values) and cognitive or epistemic values that are considered properly internal to science. Although we mean to problematize this distinction, as many others have in recent years, including Longino herself (2002), it provides a useful framework for our discussion precisely because it remains so widely influential in both popular and philosophical literature and within the sciences themselves. For an especially interesting reconfiguration of this distinction, see Solomon's characterization of empirical and nonempirical decision vectors (2001, 51–64).

3. The scope of these projects, as they had taken shape in the decade before Noz-ick gave his APA address, was outlined by Maffie (1990) and represented in contributions to Schmidt (1994) and to Megill (1994). The commitment to various forms of Nozick's "third option" was especially evident in Callebaut's interviews with naturalizing philoso-phers of science (1993) and in contributions to Kornblith (1993) and to Galison and Stump (1996).

4. The fruits of these critical and constructive labors have been described in consid-erable detail by a number of feminist practitioners and science studies scholars. Sum-maries are included in work we have published (e.g., Nelson 1990; Wylie 1997a, 1997b; Nelson and Wylie 2004), as well as in *Has Feminism Changed Science?* (Schiebinger 1999) and in contributions to a special issue of *Osiris, Women, Gender, Science* (Kohlstedt and Longino 1997), and to *Feminism in Twentieth Century Science, Technology, and Med-icine* (Creager, Lunbeck, and Schiebinger 2001).

5. In "Objectivity, Value Judgment, and Theory Choice," Kuhn argues as follows:

> Five characteristics—accuracy, consistency, scope, simplicity, and fruitfulness—are all standard criteria for evaluating the adequacy of a theory. . . . Neverthe-less, two sorts of difficulties are regularly encountered by men [*sic*] who must use these criteria in choosing [between theories] . . . Individually, the criteria are imprecise: individuals may legitimately differ about their application to concrete cases. In addition, when deployed together, they repeatedly prove to conflict with one another; accuracy may, for example, dictate the choice of one theory, scope the choice of its competitor. (Kuhn 1977, 322)

6. Consider two other examples that are widely discussed in connection with com-promising bias along race and class lines. The critics who most sharply questioned Burt's widely influential IQ studies seem to have been motivated to scrutinize the "procedural" aspects of his research because it was simply not plausible to them that working-class chil-dren should consistently prove to be less intelligent than their upper-class peers (see, e.g., Kamin and Layzer in Block and Dworkin 1976). In a similar vein, Collins has described the (constructive, critical) difference it makes to be a raced, as well as gendered, "insider outsider" to sociology: The misrepresentations of black family structure and employment patterns inherent in post-Moynihan sociology were patently obvious to her as one who brought to her professional training as a sociologist a grounding in the culture, history, and experience of the communities in question (Collins 1991).

7. Before 1973, the representation of women was never more than 13% in the Soci-ety for American Archaeology; in that year it jumped to 18%, reaching 30% by 1976 and stabilizing at 36% after 1988 (see Patterson 1995, 81–83; Wylie 1994, 65–66).

8. These cases are discussed in more detail in Wylie (2002).

9. This is shorthand for something like: the Y chromosome promotes the synthesis of the H-Y antigen, which in turn promotes the development of the embryonic testis, which in turn is able to synthesize testosterone and MIS, and the former promotes further developments of male ducts and external genitalia while the latter promotes degeneration of female ducts.

10. For example, Carlson's *Patten's Foundations of Embryology*, a widely used un-dergraduate embryology text, states: "in the absence of the H-Y antigen the gonad later

becomes transformed into an ovary" and that "the external genitalia . . . develop . . . in the female direction if the influence of testosterone is lacking" (1981, 459–61).

11. Indeed, Fausto-Sterling attributes the growing interest in investigating a positive role for estrogen to the discovery that, in some cells of the body, testosterone may be converted into estrogen—not to an interest in estrogen on its own terms (Fausto-Sterling 1987, 67).

12. So, for example, feminist biologists have criticized the emphasis on the role of androgens in research into fetal brain development, including the hypothesis that testosterone "organizes" the developing male fetal brain and serves as the basis for sexually differentiated brains and behavior in a variety of species. The roots of the organizer hypotheses lie in research in reproductive endocrinology; as early as the 1960s, some in the field had posited "male brains" and "female brains" in rats and the hypothesis has since been extended to investigations of other species (see, e.g., Bleier 1984, 1988; Longino 1990; Nelson 1996).

13. We focus here on androcentrism, but feminist biologists have identified the role of other contextual values in this and other areas of research in the life sciences. For example, feminist biologists have argued that there are relationships between sociopolitical contexts and commitments to linear, hierarchical models of biological processes. Gilbert analyzes the role and consequences of the "dominant" and "subordinate," "husband" and "wife," and "male" and "female" metaphors invoked in characterizing, respectively, the nucleus and cytoplasm in debates about their roles that dominated developmental biology in the early twentieth century (Gilbert 1988; Gilbert et al. 1988). And Bleier (1984, 1988), Hubbard (1982, 1990), and Keller (1985) analyze the commitment to "dominant" and "subordinate" entities in the "executive DNA" model. In addition, Fausto-Sterling (1993) argues that the commitment to sexual dimorphism is as much a function of contextual values as it is of biological theory.

14. Carlson's embryology text also discusses "the tendency of the body to develop along female lines in the absence of other modifying influences" (1981, 459–61).

15. Longino's detailed list of feminist community values, from which we draw the "bottom-line" maxim and requirement of epistemic provisionality, includes two broadly ontological values that she finds characteristic of feminist research. One is a preference for hypotheses that take into account individual differences among the objects of study and that allow "equal standing for different types," by contrast to theories that rank differences or conceptualize them in terms of divergence from or failure to realize a norm: a principle of ontological heterogeneity. The other is a preference for theories that treat "complex interaction as a fundamental principle of explanation," rather than presuming that there must be some dominant or controlling factor to which causal agency can be attributed: a principle of causal complexity (Longino 1994, 477). Although these ontological values may not be generalizable to feminist research in all the social and life sciences (Wylie 1994), they do certainly capture community values that inform the feminist critiques considered here and in feminist critiques of the biological determinism that animates the quest for sex differences in cognitive capacities (as a function of the androgens organizing "male brains") and in temperament (attributed to testosterone levels), among other such examples.

16. This generates the infamous "bias paradox" (Anthony 1993; Campbell 1998).

17. The recommendation we make here is closely related to Harding's argument for "strong objectivity" (1993) and has been developed, in other terms, by Bleier (1984, 203–5), Keller (1985), and Nelson (1990, chapter 7).

REFERENCES

Anthony, L. 1993. "Quine as Feminist: The Radical Import of Naturalized Epistemology," in Anthony and Witt, eds., A Mind of One's Own: Feminist Essays on Reason and Objectivity, pp. 110–53. Boulder, CO: Westview.

Bentley, G. R. 1996. "How Did Prehistoric Women Bear Man the Hunter? Reconstructing Fertility from the Archaeological Record," in R. P. Wright, ed., Gender and Archaeology, pp. 23–51. Philadelphia: University of Pennsylvania Press.

Bernick, Kathryn, ed. 1998. Hidden Dimensions: The Cultural Significance of Wetland Archaeology. Vancouver: University of British Columbia Press.

Bernstein, R. 1983. Beyond Objectivism and Relativism. Philadelphia: University of Pennsylvania Press.

Bleier, R. 1984. Science and Gender: A Critique of Biology and Its Theories on Women. New York: Pergamon.

Bleier, R., ed. 1988. Feminist Approaches to Science. New York: Pergamon.

Block, N. J., and G. Dworkin. 1976. The IQ Controversy. New York: Random House.

Brumbach, H. J., and R. Jarvenpa. 1997. "Ethnoarchaeology of Subsistence Space and Gender: A Subarctic Dene Case." American Antiquity, 62, pp. 414–36.

Brumfiel, L. M. 1991. "Weaving and Cooking: Women's Production in Aztec Mexico," in J. M. Gero and M. W. Conkey, eds., Engendering Archaeology, pp. 224–53. Oxford: Blackwell.

Callebaut, W. 1993. Taking the Naturalistic Turn, or How Real Philosophy of Science Is Done. Chicago: University of Chicago Press.

Campbell, R. 1998. Illusions of Paradox: A Feminist Epistemology Naturalized. Lanham, MD: Rowman and Littlefield.

Carlson, B. M. 1981. Patten's Foundations of Embryology. New York: McGraw Hill.

Claassen, C., and R. A. Joyce, eds. 1997. Women in Prehistory: North America and Mesoamerica. Philadelphia: University of Pennsylvania Press.

Cohen, M. R. N., and S. Bennett. 1993. "Skeletal Evidence for Sex Roles and Gender Hierarchies in Prehistory," in B. Miller, ed., Sex Roles and Gender Hierarchies, pp. 273–96. Cambridge: Cambridge University Press.

Collins, P. H. 1991. "Learning from the Outsider Within," in M. M. Fonow and J. A. Cook, eds., Beyond Methodology: Feminist Scholarship as Lived Research, pp. 35–59. Bloomington: Indiana University Press.

Conkey, M. W., and J. M. Gero. 1997. "Gender and Feminism in Archaeology." Annual Review of Anthropology, 26, pp. 411–37.

Conkey, M. W., and J. D. Spector. 1984. "Archaeology and the Study of Gender," in M. B. Schiffer, ed., Advances in Archaeological Method and Theory, vol. 7, pp. 1–38. New York: Academic Press.

Creager, A. N. H., E. Lunbeck, and L. Schiebinger, eds. 2001. Feminism in Twentieth-Century Science, Technology and Medicine. Chicago: University of Chicago Press.

Dobres, M.-A. 1995. "Gender and Prehistoric Technology." World Archaeology, 27, pp. 25–49.

Donlon, D. 1993. "Imbalance in the Sex Ratio in Collections of Australian Aboriginal Skeletal Remains," in H. duCros and L. Smith, eds., Women in Archaeology: A Feminist Critique, pp. 98–103. Canberra: Australian National University Occasional Papers.

Engelstad, E. 2004. "Another F-Word? Feminist Gender Archaeology," in T. Oestigaard, N. Afinset, and T. Saetersdal, eds., *Combining the Past and the Present Archaeological Perspectives on Society*, pp. 39–45. Oxford: Archaeopress.

Fausto-Sterling, A. 1985. *Myths of Gender: Biological Theories about Women and Men*. New York: Basic Books.

Fausto-Sterling, A. 1987. "Society Writes Biology/Biology Constructs Gender." *Daedalus* 116, pp. 61–76.

Fausto-Sterling, A. 1993. "The Five Sexes: Why Male and Female Are Not Enough," *The Sciences* March–April, pp. 20–25.

Galison, P., and D. Stump, eds. 1996. *The Disunity of Science*. Stanford, CA: Stanford University Press.

Gero, J. M. 1993. "The Social World of Prehistoric Facts: Gender and Power in Paleoindian Research," in H. duCros and L. Smith, eds., *Women in Archaeology: A Feminist Critique*, pp. 31–40. Canberra: Australian National University Occasional Papers.

Gero, J. M. 1991. "Genderlithics: Women's Roles in Stone Tool Production," in J. M. Gero and M. W. Conkey, eds., *Engendering Archaeology: Women and Prehistory*, pp. 163–93. Cambridge: Blackwell.

Gero, J. M., and M. W. Conkey, eds. 1991. *Engendering Archaeology: Women and Prehistory*. Cambridge: Blackwell.

Gifford-Gonzalez, D. 1993. "Gaps in Zooarchaeological Analyses of Butchery: Is Gender an Issue?" in J. Hudson, ed., *Bones to Behavior*, pp. 181–99. Carbondale: Southern Illinois University Press.

Gilbert, S. 1988. "Cellular Politics," in R. Rainger, K. R. Benson, and J. Maienschein, eds., *The American Development of Biology*, pp. 311–46. Philadelphia: University of Pennsylvania Press.

Gilbert, S., et al. 1988. "The Importance of Feminist Critique for Cell Biology." *Hypatia*, 3, pp. 61–76.

Gilchrest, R. 1999. *Gender and Archaeology: Contesting the Past*. New York: Routledge.

Goy, R. W., and B. S. McEwen, eds. 1980. *Sexual Differentiation of the Brain*. Cambridge, MA: MIT Press.

Haack, S. 1993. "Knowledge and Propaganda: Reflections of an Old Feminist." *Partisan Review*, Fall, pp. 556–64.

Hanen, M. P., and J. Kelley. 1992. "Gender and Archaeological Knowledge," in L. Embree, ed., *Metaarchaeology: Reflections by Archaeologists and Philosophers*, pp. 195–227. Boston: Reidel.

Haraway, D. J. 1991. "Situated Knowledges: The Science Question in Feminism and the Privilege of Partial Perspectives," in *Simians, Cyborgs, and Women: The Reinvention of Nature*, pp. 183–202. New York: Routledge.

Harding, S. 1986. *The Science Question in Feminism*. Ithaca, NY: Cornell University Press.

Harding, S. 1993. "Rethinking Standpoint Epistemology: What Is 'Strong Objectivity'?" in L. Alcoff and E. Potter, eds., *Feminist Epistemologies*, pp. 49–82. New York: Routledge.

Hastorf, C. A. 1991. "Gender, Space, and Food in Prehistory," in J. M. Gero and M. W. Conkey, eds., *Engendering Archaeology*, pp. 132–59. Oxford: Blackwell.

Hays-Gilpin, K., and D. S. Whitely, eds. 1998. *Reader in Gender Archaeology*. New York: Routledge.

Hubbard, R. 1982. "Have Only Men Evolved?" in R. Hubbard, M. Henifin, and B. Fried, eds., *Biological Woman: The Convenient Myth*, pp. 87–106. Cambridge, MA: Schenkman.

Hubbard, R. 1990. *The Politics of Women's Biology*. New Brunswick, NJ: Rutgers University Press.

Keller, E. F. 1985. *Reflections on Gender and Science*. New Haven, CT: Yale University Press.

Kohlstedt, S. G., and H. Longino, eds. 1997. *Women, Gender, and Science: New Directions*, special issue of *Osiris*, vol. 12.

Kornblith, H., ed. 1993. *Naturalizing Epistemology*, 2nd ed. Cambridge, MA: MIT Press.

Kuhn, T. 1977. "Objectivity, Value Judgment, and Theory Choice," in *The Essential Tension: Selected Studies in Scientific Tradition and Change*, pp. 320–39. Chicago: University of Chicago Press.

Lloyd, E. 1995. "Objectivity and the Double Standard for Feminist Epistemologies." *Synthese*, 104, pp. 351–81.

Longino, H. E. 1987. "Can There Be a Feminist Science?" *Hypatia*, 2, pp. 51–64.

Longino, H. E. 1990. *Science as Social Knowledge: Values and Objectivity in Scientific Inquiry*. Princeton, NJ: Princeton University Press.

Longino, H. E. 1994. "In Search of Feminist Epistemology." *Feminist Epistemology—For and Against*," special issue of *Monist*, 77, pp. 472–85.

Longino, H. E. 1995. "Gender, Politics, and the Theoretical Virtues." *Synthese* 104, pp. 383–97.

Longino, H. E. 1997. "Cognitive and Non-Cognitive Values in Science: Rethinking the Dichotomy," in M. C. Nelson and S. M. Nelson, eds., *Feminism, Science, and the Philosophy of Science*, pp. 39–58. Dordrecht: Kluwer Academic.

Longino, H. E. 2002. *The Fate of Knowledge*. Princeton, NJ: Princeton University Press.

Longino, H. E., and R. Doell. 1983. "Body, Bias, and Behavior: A Comparative Analysis of Reasoning in Two Areas of Biological Science." *Signs*, 9, pp. 206–27.

Maffie, J. 1990. "Recent Work on Naturalized Epistemology." *American Philosophical Quarterly*, 27, pp. 281–93.

McGuire, K. R., and W. R. Hildebrandt. 1994. "The Possibilities of Women and Men: Gender and the California Millingstone Horizon." *Journal of California and Great Basin Archaeology*, 16, pp. 41–59.

Megill, A., ed. 1994. *Rethinking Objectivity*. Durham, NC: Duke University Press.

Money, J. 1970. *Love and Lovesickness*. Baltimore, MD: Johns Hopkins University Press.

Nelson, L. H. 1990. *Who Knows: From Quine to a Feminist Empiricism*. Philadelphia: Temple University Press.

Nelson, L. H. 1996. "Empiricism without Dogmas," in L. H. Nelson and J. Nelson, eds., *Feminism, Science, and the Philosophy of Science*, pp. 95–120. Dordrecht: Kluwer Academic.

Nelson, L. H., and A. Wylie. 2004. "Introduction," in *Feminist Science Studies*, special issue of *Hypatia*, 19, pp. vii, xiii.

Nelson, M. C., S. M. Nelson, and A. Wylie, eds. 1994. *Equity Issues for Women in Archaeology*. Washington, DC: American Anthropological Association.

Nelson, S. M. 1997. *Gender in Archaeology: Analyzing Power and Prestige*. Walnut Creek, CA: AltaMira.

Nozick, R. 1998. "Invariance and Objectivity." *Proceedings and Addresses of the American Philosophical Association* 72.2(1998): 21–48.

Patterson, T. C. 1995. *Toward a Social History of Archaeology in the United States.* Orlando, FL: Harcourt Brace.

Sassaman, K. 1992. "Lithic Technology and the Hunter-Gatherer Sexual Division of Labor." *North American Archaeologist,* 13, pp. 249–62.

Schiebinger, L. 1999. *Has Feminism Changed Science?* Cambridge, MA: Harvard University Press.

Schmidt, F. F., ed. 1994. *Socializing Epistemology.* Lanham, MD: Rowman and Littlefield.

Slocum, S. 1975. "Woman the Gatherer: Male Bias in Anthropology," in R. Reiter, ed., *Toward an Anthropology of Women,* pp. 36–50. New York: Monthly Review Press.

Soffer, O. 2004. "Recovering Perishable Technologies through Use Wear on Tools: Preliminary Evidence for Upper Paleolithic Weaving and Net Making." *Current Anthropology* 45: 407–418.

Soffer, O., J. M. Adovasio, and D. C. Hyland. 2001. "Perishable Technologies and Invisible People: Nets, Baskets, and 'Venus' Wear ca. 26,000 b.p," in B. A. Purdy, ed., *Enduring Records: The Environmental and Cultural Heritage of Wetlands,* pp. 233–45. Oxford: Oxbow.

Solomon, M. 2001. *Social Empiricism.* Cambridge MA: MIT University Press.

Sperling, S. 1991. "Baboons with Briefcases vs. Langurs in Lipstick," in M. di Leonardo, *Gender at the Crossroads of Knowledge,* pp. 204–34. Berkeley: University of California Press.

Suppe, F., ed. 1983. *The Structure of Scientific Theories,* 2nd ed. Chicago: University of Illinois Press.

Tringham, R. E. 1995. "Archaeological Houses, Households, Housework and the Home," in D. Benjamin and D. Stea, eds., *The Home: Words, Interpretations, Meanings, and Environments,* pp. 76–107. Aldershot: Avebury.

Tuchmann-Duplessis, H., M. Aurous, and P. Haegel. 1975. *Illustrated Human Embryology.* New York: Springer-Verlag.

Walde, D., and N. D. Willows, eds. 1991. *The Archaeology of Gender* (proceedings of the 22nd annual Chacmool Conference). Calgary: Archaeological Association of the University of Calgary.

Wright, Rita P., ed. 1996. *Gender and Archaeology.* Philadelphia: University of Pennsylvania Press.

Wylie, A. 1994. "The Trouble with Numbers: Workplace Climate Issues in Archaeology," in M. C. Nelson, S. M. Nelson, and A. Wylie, eds., *Equity Issues for Women in Archaeology,* Archaeological Papers of the American Anthropological Association, Number 5, pp. 65–71. Washington: American Anthropological Association.

Wylie, A. 1995. "Doing Philosophy as a Feminist: Longino on the Search for a Feminist Epistemology." *Philosophical Topics* 23, pp. 345–58.

Wylie, A. 1997a. "Good Science, Bad Science, or Science as Usual? Feminist Critiques of Science," in L. D. Hager, ed., *Women in Human Evolution,* pp. 29–55. New York: Routledge.

Wylie, A. 1997b. "The Engendering of Archaeology: Refiguring Feminist Science Studies." *Osiris,* 12, pp. 80–99.

Wylie, A. 2000. "Rethinking Objectivity: Nozick's Neglected Third Option." *International Studies in Philosophy of Science*, 14, pp. 5–10.

Wylie, A. 2001. "Doing Social Science as a Feminist: The Engendering of Archaeology," in A. Creager, E. Lunbeck, and L. Schiebinger, eds., *Feminism in Twentieth-Century Science, Technology and Medicine*, pp. 23–45. Chicago: University of Chicago Press.

Wylie, A. 2002. "The Constitution of Archaeological Evidence: Gender Politics and Science," in A. Wylie, ed., *Thinking from Things: Essays in the Philosophy of Archaeology*, pp. 185–99. Berkeley: University of California Press.

Wylie, A. 2003. "Why Standpoint Matters," in R. Figueroa and S. Harding, eds., *Science and Other Cultures: Issues in Philosophies of Science and Technology*, pp. 26–48. New York: Routledge.

FOUR

EVALUATING SCIENTISTS: EXAMINING THE EFFECTS OF SEXISM AND NEPOTISM

K. Brad Wray

4.1 Introduction

In addition to evaluating the extent to which hypotheses and theories are supported by data, scientists are frequently required to evaluate their peers. To date, this dimension of scientific inquiry has received little attention from philosophers of science. In this chapter, I aim to examine a variety of contexts in which scientists evaluate their peers and the impact such evaluations have on science. In particular, I am interested in examining ways and contexts in which scientists' nonepistemic values influence their *epistemic* evaluations of their peers. I recommend distinguishing between what I will call "implicit" and "explicit" judgments of competence. When scientists make judgments about whose data and findings to rely on in their own research, they make *implicit* judgments of their peers' competence, whereas when they evaluate their peers' competence in the process of making funding and hiring decisions, they make *explicit* judgments. I argue that there is evidence to suggest that scientists are influenced by their nonepistemic values when they make explicit judgments and that this has an impact on their ability to realize their epistemic goals. Further, I claim that we have good reasons to believe that they are not so influenced when they make implicit judgments of their peers.

In section 4.2, I draw the distinction between implicit and explicit judgments of competence and give examples of each sort of judgment. In section 4.3, I review a number of studies that suggest scientists are influenced by nonepistemic values when they make explicit judgments of competence. In particular, scientists are affected by sexism and nepotism. In section 4.4, I examine the epistemic consequences of scientists' biased explicit judgments of competence and focus on the adverse effects of nepotism. Then, in section 4.5, I explain why scientists are not prone to the same problems when making implicit judgments of competence.

4.2 Implicit and Explicit Judgments of Competence

Before proceeding, I want to clarify what I mean by the terms *epistemic* and *nonepistemic*. Drawing a clear distinction between epistemic and nonepistemic values is not a straightforward task. In fact, Phyllis Rooney (1992) contests both the possibility and usefulness of drawing such a distinction. Nonetheless, I believe that Ernan McMullin (1984) provides a plausible way to draw such a distinction. According to McMullin, "an epistemic factor is one which the scientist would take to be a proper part of the argument he or she is making" (129). Following McMullin, I will count a value, factor, or consideration as epistemic just in case the agent influenced by it perceives it as contributing to understanding the world. For example, when scientists are evaluating competing theories, their appeals to the notions of simplicity or ontological heterogeneity, breadth of scope, consistency, empirical adequacy, and fruitfulness will all count as epistemic. When scientists are evaluating someone's testimony, such matters as the perceived reliability of the person will count as epistemic. On the other hand, any value, factor, or consideration that the agents do not think contribute to their understanding of the world will count as nonepistemic.[1] Consequently, if scientists are influenced in their decision making by such factors as gender or race, where specific genders and races are not believed to be correlated with epistemic reliability, we can assume that nonepistemic values, factors, or considerations are influencing them in their judgments and decision making.[2]

None of this precludes the now commonplace view that nonepistemic values can aid scientists in realizing their epistemic goals. Many recent studies of scientific practice suggest that nonepistemic factors such as the desire for peer recognition can play an important positive role in ensuring the success of science (see Hull 1988; Latour and Woolgar 1986). Still, many nonepistemic factors are apt to have a pernicious effect on science and frustrate scientists' abilities to realize their epistemic goals. That is, such factors frequently work at cross-purposes to epistemic factors.

In the remainder of this section, I want to draw attention to two very different sorts of contexts in which scientists are required to evaluate their peers' competence. Indeed, there are many different contexts in which scientists evaluate their peers, but I want to draw a distinction between explicit and implicit judgments of competence.

Let us begin by considering instances of explicit judgments of competence. Often, in the process of awarding research fellowships or funding, referees are asked to explicitly evaluate candidates' competence. For example, in their efforts to distribute postdoctoral fellowships, the Swedish Medical Research Council (MRC) asks referees to assign to each candidate a score between 0 and 4 as an assessment of the candidate's scientific competence. This score constitutes one of three measures used to determine each scientist's ranking (Wennerås and Wold 1997).

Other institutions use similar evaluation procedures that include an assessment of a scientist's competence. For example, of the four groups of criteria used by the National Science Foundation (NSF) in making research funding decisions, one group includes "criteria evaluating the principal or named investigator's demonstrated competence" (Cole, Rubin, and Cole 1978, 7). In making such judgments, NSF reviewers are asked to consider:

1. The scientist's training, past performance record, and estimated potential for future accomplishments
2. The scientist's demonstrated awareness of previous and alternative approaches to his problem
3. Probable adequacy of available or obtainable instrumentation and technical support (p. 159)

Unlike the MRC reviewers, NSF reviewers do not assign a separate score of competence for applicants. Rather, NSF reviewers are required to assign only a single score to each applicant, giving consideration to the proposed research program and the competence of the principal investigator, among other things.

These examples are instances of explicit judgments of competence. Such judgments occur in other contexts also. For example, hiring committees at universities also must make assessments of the candidates' competence.

In addition, there are other contexts in which scientists must evaluate the competence of their peers *implicitly*. In the process of doing science, scientists will inevitably make judgments about whose results can and will be trusted, that is, who is reliable and whose results should be treated with suspicion, perhaps even checked. In their detailed anthropological study of the Salk Institute, Latour and Woolgar (1986) provide a clear example of scientists making such judgments. Latour and Woolgar recount the interactions of two scientists, Smith and Wilson, deliberating about whether they should rely on findings reported by some other researchers (158–59). Smith and Wilson discuss the concerns they have with the other researchers' findings and methods, as well as the implications their findings have for their own research.

Steven Shapin (1994) claims that scientists have been deeply dependent on others in their research efforts since the early days of the scientific revolution. John Hardwig (1991) argues that the increasing specialization characteristic of contemporary science has made scientists even more dependent on their peers than ever before. As Hardwig (p. 693) explains, "Modern knowers cannot be independent and self-reliant, not even in their own fields of specialization." Hardwig (p. 695) cites two factors that have contributed to making scientists more deeply dependent on their peers than ever before. First, "the process of gathering and analyzing data sometimes just takes too long to be accomplished by one person." As Hardwig notes, "The pace of science is often far too rapid for a lone experimenter to make any contribution at all by [gathering and analyzing data individually]." Second, "research is increasingly done by teams because no one knows enough to be able to do the experiment by herself." For example, Hardwig (1985) cites an article reporting the life span of charmed particles that was coauthored by ninety-nine scientists from a variety of specializations in physics.[3]

When scientists decide to rely on others' findings, they must trust that what others report happened did in fact happen. And underlying one's decision to trust another is a tacit judgment of the other's competence. As Hardwig (1991, 700) explains, the reliability of a person's testimony depends on the reliability of their character; consequently, as

Hardwig puts the point, "scientific knowledge rests on trust . . . in the character of scientists" (p. 707). It seems that scientists are now unable to engage in science at all without making implicit judgments of other scientists' competence.

4.3 The Influence of Nonepistemic Values on Explicit Judgments of Competence

In this section, I have two goals. First, I examine evidence from a study that suggests that scientists are not as influenced by epistemic factors as we would expect them to be when they evaluate their peers. Second, I examine two studies that aim to identify some of the nonepistemic factors that do influence scientists when they evaluate their peers.

Two of these studies I examine rely on citations as a measure of quality. I recognize that some sociologists have questioned whether citations really reflect the epistemic merits of scientific research. But, as Jonathan Cole (1979, 16) notes, "While citations are hardly an ideal measure of quality, their frequency turns out to be strongly correlated with many other independent appraisals of the quality of research performance, such as prestige of honorific awards held by scientists, and peer evaluation of the quality of research contributions." Further, "in the absence of counts of citations, the number of publications does reasonably well as a measure of research performance. There is a strong although not perfect association between citation counts and number of publications." As imperfect as such measures might be, they may be the best measures we have.[4]

In their study of peer review in the NSF, Cole, Rubin, and Cole (1978) found evidence that suggests that scientists evaluating their peers were not influenced by the sorts of factors that one might regard as epistemically relevant. Cole, Rubin, and Cole assumed that a scientist's competence should be reflected by measures of the following sort:

1. The quantity of frequently cited papers
2. The quantity of well-known work published more than ten years ago
3. Whether one was a past recipient of an NSF grant

They reasoned that scientists who had a great number of frequently cited papers and a great quantity of well-known work published more than ten years ago and were past recipients of the competitively sought NSF grants should generally get higher ratings from NSF reviewers. The higher ratings would reflect the scientists' greater competence.

However, Cole, Rubin, and Cole found only a weak correlation between the fact that scientists had demonstrated "their competence by publishing frequently cited papers" and their receiving favorable ratings by the NSF referees for their grant proposals. As they note, "a substantial portion of the ratings of scientists with relatively large numbers of citations are relatively [low]" (p. 55). In fact, dividing the applicants into quintiles according to how frequently they are cited, they found that "in fluid mechanics, [for] those in the first four quintiles, there is no difference whatsoever in the proportion receiving excellent or very good ratings" (p. 62). Similarly, in anthropology, scientists in the highest citation category and those in the lowest are equally likely to get a rating of excellent or very good (p. 60). Further, they found that "scientists who were well known as a result of work published more than 10 years ago are only slightly more likely to get higher ratings than scientists who are not well known on the basis of work published 10 or more years ago" (pp. 62–63). In addition, they found that "whether or not applicants are recent past recipients of NSF funds has very little influence on the ratings of their current applications" (p. 67). Cole, Rubin, and Cole conclude that "if reviewers are being influenced at all by the past performances and reputations of principal investigators, the influence is not very great" (p. 65). Hence, in evaluating grant applicants, scientists are less influenced by genuinely relevant epistemic considerations than they should be.[5]

Now I want to examine evidence that will enable us to identify the sorts of nonepistemic factors that are influencing scientists when they are making explicit judgments of their peers' competence. The influence of identifiable nonepistemic values on scientists' explicit judgments of their peers is most evident in Wennerås and Wold's study of the MRC (1997). They found that "peer reviewers overestimated the male achievements and/or underestimated female performance" and "the peer reviewers deemed women applicants to be particularly deficient in scientific competence" (p. 341). Wennerås and Wold demonstrated this by comparing the scores of competence assigned by the MRC to a measure of competence that accounted for the quantity and quality of each candidate's publications, given that "it is generally regarded that [competence] is related to the number and quality of scientific publications" (p. 341).[6] The measure of quality was determined by the "impact factor" of the journals in which the articles were published, and the precise value for each journal was determined, not by Wennerås and Wold, but by an independent body, the Institute for Scientific Information, the institution responsible for the production of the *Science Citation Index*. The impact factor for a journal is determined by the number of citations in journals indexed by the *Science Citation Index* to articles published in that journal, divided by the number of articles published in that journal in a year (Garfield 1989, 38A).

Wennerås and Wold found that "men and women of equal scientific productivity [did not] receive the same competence rating by the MRC reviewers"; "the peer reviewers gave female applicants lower scores than male applicants who displayed the same level of scientific productivity" (p. 342). In fact, Wennerås and Wold found that when the candidates were divided into five equal groups according to their own method of determining competence scores, "the most productive group of female applicants . . . was the only group of women judged to be as competent as men, although only as competent as the least productive group of male applicants" (p. 342).

Wennerås and Wold found that two factors besides scientific competence had an impact on the competence scores assigned by the referees: "the gender of the applicant and the affiliation of the applicant with a committee member" (p. 342). Wennerås and Wold found that "a female applicant had to be 2.5 times more productive than the average male applicant to receive the same competence score as he" (p. 342). And applicants who were affiliated with a committee member had an advantage similar to male applicants, even though committee members were not allowed to evaluate applications from applicants with whom they had some affiliation.[7]

Wennerås and Wold's study is not the only one that raises concerns about nepotism. Prompted by Congress, the General Accounting Office (GAO) initiated a study of the peer review processes used by NSF, the National Institutes of Health, and the National Endowment for the Humanities. The GAO found that "at NSF . . . 'a reviewer's personal familiarity with an applicant was associated with better scores' in the review process," as well as "'empirical evidence of potential problems'—including gender and racial bias" (Marshall 1994, 863).

Another study, conducted by Steinpreis, Anders, and Ritzke (1999), found that similar biases influence psychologists' assessments of job candidates for university positions. Assessments of job candidates are similar in important respects to the assessments referees make on behalf of funding agencies. In particular, scientists are initially required to assess other scientists as a whole, on the basis of their curriculum vitae (c.v.) and other supporting material, rather than by assessing a particular argument or paper. In their study, Steinpreis, Anders, and Ritzke distributed a c.v. of a real job candidate to a number of randomly selected male and female psychologists working at universities. Half of the subjects received a c.v. on which a typical female name appeared, and the other half received an identical c.v. on which a typical male name appeared. When the questionnaire asked participants whether they would hire the applicant, "participants were more likely to hire the male applicant than the female applicant" (p. 520).[8] Further, the researchers found that "both males and females demonstrated the same gender bias in favor of male applicants" (p. 520).[9]

To sum up, it appears that nonepistemic values and considerations influence scientists in their explicit evaluations of their peers. In particular, sometimes genuine epistemic qualities are just disregarded, and on other occasions, scientists are influenced in their judgments by functionally irrelevant factors, like gender and network ties. This, I believe, should lead us to question whether the content of science is adversely affected as a result.

4.4 The Epistemic Consequences of Nepotism

In this section, I want to draw attention to some of the epistemic consequences of the fact that scientists' nonepistemic values influence their explicit evaluations of their peers. My analysis focuses on the problems resulting from nepotism.[10]

The principal worry is that novel lines of research may be abandoned prematurely. That is, when such functionally irrelevant factors as network ties influence who is awarded funding, then research communities are apt to be more homogeneous than they could or should be. There are at least three ways this could have a negative impact on scientists' prospects of realizing their epistemic goals.

First, scientists are apt to be slower in finding a solution to the problems that motivate their research. Philip Kitcher (1993, 70) raises this concern and argues that "the differences among scientists are not accidental but essential to continued growth: the development of [a] field would be stunted if uniformity were imposed." As Kitcher explains, a scientific research community as a whole is apt to be more effective at realizing its epistemic goals if it is characterized by cognitive diversity, with different scientists employing different techniques, working within different theoretical frameworks, and employing different inferential strategies. When there is cognitive diversity, scientists explore more of the possible solutions. Further, scientists working in a cognitively diverse research community can share information, which may assist others in not pursuing approaches to problems that have already proved to be unpromising. Cognitive diversity thus ensures that scientists are more apt to discover the truth of the matter under investigation.

But cognitive diversity can and needs to be fostered, and funding agencies should be aware of their power to both encourage and discourage cognitive diversity. Kitcher claims that his modeling of the cognitive division of labor in science suggests that "the *optimal* decision for [a scientific] community involves a much more pronounced division of labour . . . than is found in most scientific communities" (p. 360). When one of the factors that influences who is awarded a postdoctoral fellowship is whether the candidate is affiliated with a referee, as Wennerås and Wold suggest, then the research projects that are apt to be funded are more likely to be consonant with the views of the referees. Indeed, in a world in which funding for science is limited, there will be limits to the number of hypotheses that we can expect to be developed. But given the biases that currently influence funding committees, the variety and range of hypotheses developed is probably even narrower than what is financially feasible.

Second, nepotism is apt to lead to conditions that make it more difficult for scientists to adequately evaluate the existing well-developed competing hypotheses. Given the comparative nature of theory choice and evaluation, alternative hypotheses are required to ensure that scientists are able to adequately scrutinize hypotheses in a thorough manner (Okruhlik 1998, 201). In fact, as Kathleen Okruhlik notes, even false hypotheses play a constructive role in science by providing a richer comparison class from which to make assessments. Consequently, efforts should be made to ensure that alternative hypotheses are developed and that those willing to develop them are funded, especially when those willing to develop them have proven their competence by producing work that is published in leading journals. Given the nepotism that Wenerås and Wold have unearthed, it seems that in allowing their nonepistemic values to influence their judgments, scientists are making their jobs more difficult. When scientists are choosing between competing theories and hypotheses, they are apt to have fewer well-developed alternatives to choose from than they should or could have. This relative paucity reduces their ability to scrutinize the available well-developed alternatives. As a result, when a consensus does form, it may form prematurely or, alternatively, dissensus may persist longer than it needs to.

Third, when considerations of network ties influence scientists' evaluation of their peers, certain background assumptions are apt to become entrenched and shielded from critical scrutiny. This, too, can impede the growth of science. Unless a wide range of research is supported, scientists' background assumptions are less likely to be exposed to critical scrutiny. As Helen Longino (1990) argues, alternative hypotheses and research programs play the vital role of exposing background assumptions so that they can be subjected to critical scrutiny. For example, Okruhlik (1998, 204) argues that "because of the pervasiveness of gender ideology in our culture, [sexist background] assumptions generally are not called into question and are sometimes not even noticed. It is usually the case that they come to light only in the presence of a rival hypothesis." Much feminist science and philosophy of science has been devoted to playing this role, exposing otherwise invisible sexist background assumptions. For example, Sarah Hrdy's (1999) *The Woman That Never Evolved* makes explicit that the sexist assumptions of early primatologists prevented them from seeing the competitiveness of female primates. Her detailed study makes otherwise unnoticed background assumptions visible, whether or not the project of sociobiology that underwrites her research is acceptable.

Similar considerations could be raised with respect to diversity of methods. That is, nepotism not only limits the number and variety of hypotheses developed but also is apt to limit the number and variety of methods employed, which, in turn, can impede scientists' abilities to realize their epistemic goals. Methodological pluralism is probably as important to the effective operation of scientific research communities as is theoretical pluralism.

Scientists who are not awarded research fellowships or grants are not necessarily prevented from pursuing their research projects, but doing the research is apt to be more difficult without the funding. Alternative sources of funding may carry with them greater obligations that reduce the time they have to spend on their research and developing their views. Indeed, in a recent study of biological research in Australia, Bourke and Butler (1999, 498) confirm that "researchers appointed to full-time research positions . . . achieve higher visibility for their research than do researchers with significant other duties (such as teaching or clinical work)." Full-time researchers are afforded opportunities to identify and work on "research problems of wider and deeper content, closer to 'state of the art' work in the field and accordingly [are] more likely to achieve publication in the main international journals" (p. 499). Further, "research that is not funded may never take place at all" (Wessely 1996, 885), for, as Stephen Cole (1992, 85) notes, "without financial support it is frequently impossible to develop new ideas." It would be epistemically detrimental if research like Hrdy's, for example, had not been done merely because she did not get the funding that would permit her the time needed effectively to execute her project.

In summary, because nepotism is apt to limit the range of hypotheses developed in a research community, when scientists are affected by nepotism, the pursuit of their epistemic goals is adversely affected in at least three ways:

1. A scientific community's prospects of arriving at the truth is reduced.
2. A scientific community's ability to effectively evaluate competing hypotheses is reduced.
3. Scientists' background assumptions are apt to be rendered more opaque and thus less likely to be critically scrutinized.

I want to emphasize that network ties do not just have a negative epistemic impact on science. In fact, network ties serve an important epistemic function in science. Often, it is through network ties that valuable unpublished information is transmitted throughout a scientific research community. Diana Crane (1972, 25) has found that scientific research communities that lack strong network ties are actually less productive.

Further, David Hull (1988) has suggested that small research groups provide an important forum in which scientists are able to test new ideas.[11] Consequently, any efforts to reduce the negative impact that network ties have on science should be taken in such a manner so as to minimize damage to the informal communication network.

4.5 Implicit Judgments of Competence

In light of the concerns discussed earlier that suggest scientists are frequently influenced by their nonepistemic values when they evaluate their peers, it may seem that we should be equally concerned in contexts where scientists are required to implicitly evaluate their peers. Unless we have good grounds for thinking otherwise, it seems reasonable to assume that scientists' implicit judgments of their peers are as influenced by their nonepistemic values as are their explicit judgments. In this section, I want to review data that may aid us in determining whether implicit judgments of competence are also biased. I argue that the evidence suggests that scientists are not as apt to be biased in their judgments when they implicitly evaluate their peers' competence. Further, I provide two reasons that scientists are less apt to be influenced by their nonepistemic values when implicitly evaluating their peers than they are when they explicitly evaluate them. Here, my investigation focuses on gender bias rather than nepotism.

A belief common to both those offering evidence that suggests that scientists are biased in their implicit judgments and those offering evidence that suggests that scientists are not biased in their implicit judgments is that men and women make equally capable scientists and their contributions are of equal value. Consequently, given this assumption, if there is no bias, then articles authored by men and women scientists should be cited equally. Thus, if we compare the rates of citations to papers authored by men with those to papers authored by women, then we can ascertain whether scientists are biased in their implicit evaluations of their peers.

Some studies seem to suggest that scientists are biased in their implicit judgments. For example, in a study of paleoindian research, Joan Gero (1993, 37) claims that "women in paleo studies . . . are cited much less frequently than their male counterparts." In support of this claim,

Gero notes that Eileen Johnson, an important woman North American paleo field researcher, "has been cited only two and three times respectively for her singly authored pieces in the 1978 and 1980 volumes of the *Plains Anthropologist*. In contrast, comparable site data reported in the same journal by Dennis Stanford in 1978 and by Joe Ben Wheat in 1979, are cited eight and nine times respectively" (pp. 37–38). Gero adds that "Eileen Johnson's two co-authored pieces, with [male] Vance Holliday (in the same journal), dated 1980 and 1981, are cited eleven times and seven times . . . [and her] non-Paleoindian publications . . . are frequently cited" (p. 38).

Also, Marianne Ferber (1986, 384) found that, in economics, "works by women constitute a larger proportion of the citations in articles written by women than in those written by men . . . [and] similarly, works by men constitute a significantly larger proportion of the citations in articles by men than in those by women." Given these citation patterns, Ferber believes that "women are at a disadvantage in accumulating citations in a field such as economics, in which they constitute a minority" (p. 382). Consequently, Ferber argues, citations "should not be regarded as unbiased indicators of merit" (p. 389). Ferber seems to imply that, given the gendered citation patterns she discovered, we should expect that, on average, articles authored by women are cited less often than articles authored by men in fields where men outnumber women.

Unfortunately, there are methodological concerns surrounding the interpretation of both sets of data. Gero's sample is too small to enable us to draw meaningful conclusions, and there is reason to believe that Ferber's sample may not be representative, for she notes that she eliminated from her sample "those articles with no citations" (p. 383). Including the uncited papers in her study may have had a significant impact on the conclusions she drew.

Let us now consider the evidence that suggests that scientists are not biased when they make implicit evaluations of their peers. In a study of scientific productivity in chemistry, Barbara Reskin (1978, 1236) found that men are cited more frequently than women, but that once "the differences for the measures of publication quality . . . were adjusted for quantity," then the differences in citations were not statistically significant. Reskin's sample consisted of 229 females and 221 males drawn from "chemists who obtained their Ph.D.s at U.S. universities between 1955 and 1961" (p. 1235).

Similar results were found by Cole and Zuckerman (1984) in their study of scientists in six different fields. Indeed, they note that "more than 50 studies of scientists in various fields show that women publish less than men," and "the correlations between gender and productivity have been roughly constant since the 1920s" (p. 217). But they found that "although men are cited more often overall, it turns out that there are no gender differences in average citations per paper" (p. 235). More precisely, they found that "women's papers averaged 5.02 citations and men's, 4.92 (p=n.s.)" (p. 235). Cole and Zuckerman's study involved "263 matched pairs of men and women scientists (N = 526) who received doctorates in 1969–1970" (p. 217).[12]

Finally, Ray Over (1990, 333) conducted a study that compared "the sex ratio among authors of frequently cited articles with the sex ratio among authors of infrequently cited articles published in the same journal." The focus of his study was the discipline of psychology, and he examined "564 matched pairs of articles" (p. 335). Over found that "one sex was not significantly more likely than the other sex to have produced high-impact articles rather than low-impact articles" (p. 336). More precisely, he found that "in the case of single-author publications, the majority (78.0%) of high impact articles had been generated by men, but so had the majority (83.2%) of low-impact articles" (p. 335). And in the case of multiple-author publications, "men were the first-listed author of 80.3% of high-impact articles, but they were also the first-listed author of the majority (83.8%) of the low-impact articles" (p. 336).

Thus, we have reason to believe that when scientists implicitly judge their peers' competence in deciding whose results can be trusted, they are less apt to be influenced by their nonepistemic values. In the remainder of this section, I offer an explanation for why this might be so, given that scientists are biased in their explicit evaluations of their peers. There are two considerations that may account for this.

First, it seems that scientists are apt to be less biased in their implicit judgments because it is in their personal *interest* to be less biased. As Hull (1988, 4) explains, "science is so structured that scientists must, to further their own research, use the work of other scientists." Consequently, when scientists are making decisions about whose results to rely on in their efforts to advance their own research, they have an interest in using *anyone's* results that can strengthen their case. As a result, they are less apt to disregard published findings that are useful for their project merely because they are reported in an article authored by a fe-male scientist than they are to misjudge a female scientist's competence as part of a funding or hiring decision. The former sort of decision has a direct and immediate impact on scientists' own research, for they may be unable to make a particular claim unless they cite particular find-ings. The latter sort of decision, on the other hand, is apt to have only an indirect impact on a scientist's own research, if it has an impact at all.

Second, it seems that scientists have less opportunity to allow their decision making and judgments in these contexts to be influenced by nonepistemic factors and values. Nonepistemic factors and values seem to influence scientists most when they are confronted with choices. As Stephen Cole (1992) explains, scientists' judgments are more influ-enced by functionally irrelevant criteria, like a consideration of network ties that leads to nepotism, when the impersonal and objective criteria are unable to single out a superior candidate. In a variety of contexts, objective criteria do not single out a superior candidate, and scientists are confronted with choices. For example, Nobel Prize committees are confronted with a surplus of eligible and deserving candidates (p. 188). Hiring committees at colleges and universities face a similar surplus (p. 186), as do many funding agencies (p. 204). In such circumstances, there are just too many deserving and able candidates, and it is in these sorts of circumstances that nonepistemic considerations are apt to have an influence.

Things are quite different, though, when it comes to making judg-ments about whose results to rely on. A scientist is usually in need of something quite specific. As a result, scientists seldom find themselves in a position similar to the hiring committee, with numerous different but equally valuable sources reporting the sorts of findings they need to support the claims they want to make in their own research. Conse-quently, scientists are less apt to be influenced by their nonepistemic values in making such judgments.

4.6 Concluding Remarks

In summary, I have suggested that scientists' nonepistemic values influ-ence not only their judgments of hypotheses and theories but also their judgments of their peers. I have shown that when scientists' nonepistemic values influence their assessments of other scientists, scientists are apt to be frustrated in their pursuit of their epistemic goals. I have also provided grounds for believing that when scientists make implicit judgments of their peers' competence in deciding whose results can be trusted, they are less apt to be influenced by their nonepistemic values. Their interests in advancing their own research plays an important role in ensuring that such values do not have the same sort of impact on these judgments as they appear to have in hiring and funding decisions.

I suspect that part of the reason for this asymmetry is that implicit and explicit judgments of competence have an impact on scientific inquiry at different stages. Scientists make implicit judgments when deciding whose research to cite in their efforts to advance their own research. Such judgments contribute to the justification of research. Scientists make explicit judgments of competence when they are making funding and hiring decisions. These judgments have an impact on the variety of hypotheses developed and thus influence the discovery stage. Because explicit judgments of competence affect science by limiting the number of well-developed hypotheses that scientists have to choose between, I suspect that their impact on science is frequently underestimated. Scientists do not appreciate the value of what they never get to see — undeveloped hypotheses. I have argued, though, that there is good reason to believe that when biases limit the number of well-developed hypotheses in a research community, scientists are apt to be less able to realize their epistemic goals.

When scientists make implicit judgments of their peers, their own epistemic and career interests ensure that they are less biased than they might otherwise be. But in the various contexts where they are required to make explicit judgments of their peers' competence, their decisions have little impact on their subsequent ability to realize their personal epistemic goals. Ironically, it is when scientists have strong personal interests in the outcome of their decisions that they are most inclined to make unbiased evaluations of their peers' competence. Their personal interests motivate them to act in ways beneficial to the scientific research community as a whole.

It seems that if we are going to eliminate the influence of scientists' nonepistemic values on their explicit judgments of their peers, we need some counterbalance like the interests that are at work when scientists make implicit judgments. One possibility is for scientists to operationalize the notion of competence when they have to make explicit judgments of their peers' competence, as Wennerås and Wold do in their study. By doing this, we are more apt to ensure that functionally irrelevant selection criteria do not have the impact on science that they currently appear to have. Further, it is probably best that, like the MRC and unlike the NSF, scientists assign a separate score to their peers' competence when making funding decisions. This will aid scientists in reducing the impact of their nonepistemic values on funding decisions.

ACKNOWLEDGMENTS

I thank the following people for valuable feedback on earlier drafts: Lori Nash, Kristina Rolin, Janet Kourany, David Hull, John Hardwig, Rob Hudson, Jonathan Cohen, and Keith Douglas. Kristina Rolin's (2002) "Gender and Trust in Science" was very useful as I worked out my own position on the issues discussed here. I thank the editors for their comments, which enabled me to improve my argument in section 5. I thank Harold Kincaid for inviting me to participate in the "Value-Free Science: Ideal or Illusion?" conference, and the Center for Ethics and Values in the Sciences at the University of Alabama at Birmingham for their generous support. I also thank my audiences at the "Value-Free Science" conference and the Canadian Philosophical Association's annual meeting, to whom I presented earlier drafts. Finally, I thank the Canadian Philosophical Association for their generosity in paying my airfare to Quebec City.

NOTES

1. McMullin's (1984) account is historically sensitive, acknowledging that certain factors can be epistemic at one time and yet come to be nonepistemic at another time. For example, as McMullin explains, in the early days of the scientific revolution, certain metaphysical and religious considerations that are now regarded as nonepistemic were then widely regarded by scientists as epistemic.

2. As I have construed the notion of an epistemic factor, the sorts of judgments that led Robert Boyle and his peers to treat the testimony of women and servants cautiously would count as epistemic. As Steven Shapin (1994) reports, Boyle and the other members of the Royal Society *believed* that women and servants were prone to lie for gain in virtue of their constrained and dependent social and economic conditions. Gentlemen, on the other hand, did not have such incentives to lie for gain.

3. Paul Thagard (1997, 255) also suggests that "collaboration was essential for the development of the bacterial theory of ulcers because of the involvement of several different medical specialties."

4. Strictly speaking, Wennerås and Wold do not rely on citation counts as a measure of quality. Rather, they use the impact factor of the journals in which articles were published as a measure of quality. There are advantages to this approach. Though most highly

cited articles are highly cited because of the positive contribution they make, sometimes articles are cited because they are especially problematic. A journal's impact factor reflects the quality of the average article in a journal.

5. Rob Hudson reminded me that Stephen Cole (1992, 150) uses this same data to argue that "there is very little evidence that the old boys' club hypothesis is correct." As Cole explains, "an analysis of variance showed that of the ten fields [they studied], only in biochemistry are reviewers from top-ranked departments more likely to give favorable ratings to applicants from top-ranked departments than would be expected by chance."

6. Wennerås and Wold devised six different measures, but they found that the various measures yielded similar results, and they reported their findings in terms of the measure I discuss here.

7. Unfortunately, Wennerås and Wold do not explain explicitly what constitutes a personal tie; I assume that it would include being in either the student-supervisor, the postdoctoral-sponsor, or the laboratory assistant–director relationship.

8. This study is more involved than I report here. They also had scientists evaluate the c.v. of the same candidate when she or he was applying for tenure, and whether the applicant should be awarded tenure. These dimensions of their study, though, are extraneous for my purposes here.

9. Wennerås and Wold were unable to determine if male and female referees for the MRC were equally biased because their sample size of female referees was too small.

10. Until we have a better sense of how gender correlates with methodological and theoretical presuppositions, I suspect that it will be difficult to assess the impact that sexist explicit judgments of competence have on scientists' abilities to realize their epistemic goals.

11. In my "The Epistemic Significance of Collaborative Research," I argue that there are other, often unnoticed, epistemic benefits that are a consequence of the social relations between scientists (Wray 2002). In particular, collaborative research affords a variety of epistemic benefits that are seldom fully recognized.

12. Long (1992, 174) also reaches a similar conclusion. As Long explains, female scientists' "published work is not marginal—it is used and cited by others."

REFERENCES

Bourke, P., and L. Butler. 1999. "The Efficacy of Different Modes of Funding Research: Perspectives from Australian Data on the Biological Sciences." *Research Policy*, 28, pp. 489–99.

Cole, J. R. 1979. *Fair Science: Women in the Scientific Community*. London: Free Press.

Cole, J. R., S. Cole, and the Committee on Science and Public Policy. 1981. *Peer Review in the National Science Foundation: Phase Two of a Study*. Washington, DC: National Academy of Science.

Cole, J. R., and H. Zuckerman. 1984. "The Productivity Puzzle: Persistence and Change in Patterns of Publication of Men and Women Scientists," in M. J. Steinkamp and M. L. Maehr, eds., *Advances in Motivation and Achievement: Women in Science*, pp. 217–58. Greenwich: JAI.

Cole, S. 1992. *Making Science: Between Nature and Society*. Cambridge, MA: Harvard University Press.

Cole, S., L. Rubin, and J. R. Cole. 1978. *Peer Review in the National Science Foundation: Phase One of a Study*. Washington, DC: National Academy of Science.

Crane, D. 1972. *Invisible Colleges: Diffusion of Knowledge in Scientific Communities*. Chicago: University of Chicago Press.

Ferber, M. A. 1986. "Citations: Are They an Objective Measure of Scholarly Merit?" *Signs: Journal of Women in Culture and Society*, 11, pp. 381–89.

Garfield, E. 1989. "Citation Analysis as a Tool in Journal Evaluation," in E. Garfield, ed., *SCI Journal Citation Reports: A Bibliometric Analysis of Science Journals in the ISI Data Base*, 34A–41A. Philadelphia: Institute for Scientific Information.

Gero, J. 1993. "The Social World of Prehistoric Facts: Gender and Power in Paleoindian Research," in H. du Cros and L. Smith, eds., *Women in Archaeology: A Feminist Critique*, pp. 31–40. Canberra: Australian National University.

Hardwig, J. 1985. "Epistemic Dependence." *Journal of Philosophy*, 82, pp. 335–49.

Hardwig, J. 1991. "The Role of Trust in Knowledge." *Journal of Philosophy*, 88, pp. 693–708.

Hrdy, S. B. [1981] 1999. *The Woman That Never Evolved*. Cambridge, MA: Harvard University Press.

Hull, D. L. 1988. *Science as a Process: An Evolutionary Account of the Social and Conceptual Development of Science*. Chicago: University of Chicago Press.

Kitcher, P. 1993. *The Advancement of Science: Science without Legend, Objectivity without Illusion*. Oxford: Oxford University Press.

Latour, B., and S. Woolgar. [1979] 1986. *Laboratory Life: The Construction of Scientific Facts*, 2nd ed. Princeton, NJ: Princeton University Press.

Long, J. S. 1992. "Measures of Sex Differences in Scientific Productivity." *Social Forces*, 71, pp. 159–78.

Longino, H. E. 1990. *Science as Social Knowledge: Values and Objectivity in Scientific Inquiry*. Princeton, NJ: Princeton University Press.

Longino, H. E. 1994. "The Fate of Knowledge in Social Theories of Science," in F. F. Schmitt, *Socializing Epistemology: The Social Dimensions of Knowledge*, pp. 135–57. Lanham, MD: Rowman and Littlefield Publishers.

Marshall, E. 1994. "Congress Finds Little Bias in System." *Science*, 265, p. 863.

McMullin, E. 1984. "The Rational and the Social in the History of Science," in J. R. Brown, ed., *Scientific Rationality: The Sociological Turn*, pp. 127–63. Dordrecht: D. Reidel.

Okruhlik, K. 1998. "Gender and the Biological Sciences," in M. Curd and J. A. Cover, eds., *Philosophy of Science: The Central Issues*, 192–208. New York: W. W. Norton.

Over, R. 1990. "The Scholarly Impact of Articles Published by Men and Women in Psychology Journals." *Scientometrics*, 18, pp. 331–40.

Reskin, B. 1978. "Scientific Productivity, Sex, and Location in the Institution of Science." *American Journal of Sociology*, 83, pp. 1235–43.

Rolin, K. 2002. "Gender and Trust in Science." *Hypatia*, 17, pp. 95–118.

Rooney, P. 1992. "On Values in Science: Is the Epistemic/Non-epistemic Distinction Useful?" *PSA 1992*, 1, pp. 13–22.

Shapin, S. 1994. *A Social History of Truth: Civility and Science in Seventeenth-Century England*. Chicago: University of Chicago Press.

Steinpreis, R. E., K. A. Anders, and D. Ritzke. 1999. "The Impact of Gender on the Review of the Curricula Vitae of Job Applicants and Tenure Candidates: A National Empirical Study." *Sex Roles*, 41, pp. 509–28.

Thagard, P. 1997. "Collaborative Knowledge." *Nous*, 31, pp. 242–61.

Wennerås, C., and A. Wold. 1997. "Nepotism and Sexism in Peer-Review." *Nature*, 387, pp. 341–43.

Wessely, S. 1996. "What Do We Know about Peer Review?" *Psychological Medicine*, 26, pp. 883–86.

Wray, K. B. 2002. "The Epistemic Significance of Collaborative Research." *Philosophy of Science*, 69, pp. 150–68.

Zuckerman, H. [1977] 1996. *Scientific Elite: Nobel Laureates in the United States*. New Brunswick, NJ: Transaction.

PART II

EVIDENCE AND VALUES

FIVE

EVIDENCE AND VALUE FREEDOM

Elliott Sober

IN THESE POSTPOSITIVIST TIMES, THE SLOGAN "SCIENCE IS VALUE FREE" is frequently rejected disdainfully as a vestige of a bygone age. But, as often happens with babies and their bathwater, there may be something worthwhile in this slogan that we should try to identify and retain. To this end, I'll begin with two absurdities:

(1)   Scientists aren't influenced by their ethical and political values when they do science.

(2)   Scientific inference is independent of values.

I don't know if any philosopher ever believed either of these propositions. Proposition (1) is false for the simple reason that scientists are people, just like the rest of us. Perhaps they often *strive* to leave their ethical and political values at the laboratory door, but who ever thought that all of them have this aim, and that those who do succeed 100% of the time? This, by the way, does not mean that we get to assume that scientific activity can be explained *solely* in terms of the ethical and political values that scientists have. Rather, recognizing the absurdity of (1) should lead us to approach such psychological and sociological questions on a case-by-case basis. Scientists may vary among themselves, and a single scientist may be more influenced by these values in some contexts than in others. Maybe *some* scientific work proceeds completely independently of these values and *other* parts of science are entirely driven by them, and perhaps a good deal of the real world falls somewhere between these two extremes.[1]

Proposition (2) is absurd because scientific inference is regulated by normative rules. Scientists try to construct *good* tests of their hypotheses, they judge some explanations *good* and others *bad*, and they say that some inferences are *flawed* or *weak* and others are *strong*. The words I have italicized indicate that scientists are immersed in tasks of evaluation. They impose their norms on the ideational entities they construct. However, the obvious falsehood of (2) leaves it open that a restricted version of that proposition might be on the right track:

> (3)   The fact that believing a proposition would have good or bad ethical or political consequences is not evidence for or against that proposition's being true.

Is this proposition, or some refinement of it, the kernel of truth in the frequently misstated idea that "science is value free"?

We should not accept proposition (3) just because it "sounds right." After all, the evidence relation often connects facts that seem at first glance to be utterly unrelated. The proposition that the dinosaurs went extinct 65 million years ago because of a meteor hit and the proposition that there now is an iridium layer in certain rocks may appear to have nothing to do with each other. How could the presence of iridium in present-day rocks bear on the question of why the dinosaurs went extinct so long ago? Well, appearances to the contrary, there may well be such a connection (Alvarez and Asaro 1990). Why, then, should we be so sure that the ethical consequences of believing a proposition have no bearing on whether the proposition is true? This is a good question, and in the absence of a good answer, we should not complacently assume that (3) is correct.

Sometimes people believe (3) because they think there are no ethical truths in the first place. If there are no ethical truths, then ethical truths don't provide evidence for anything. Whether or not nonfactualism is correct, I think it fruitful to consider proposition (3) on the assumption that there are ethical facts. In our everyday lives, we treat ethical statements as if some of them are true. Perhaps this is a mistake, but for present purposes, let's take that practice at face value. If some normative ethical statements *are* true, what is to prevent them from standing in evidential relations with nonethical, scientific propositions?

To investigate this question, let's begin with a useful example of the evidence relation at work—smoke is evidence of fire. This relation holds because the probability of fire if smoke is present exceeds the probability of fire if smoke is absent. The two events ± smoke and ± fire are *correlated*. This is a perfectly objective relation that obtains between smoke and fire; it obtains whether or not anyone believes that it does. Understood in this way, the evidence relation has an interesting property: It is *symmetric*. If smoke is evidence of fire, then fire is evidence of smoke. This means that proposition (3) entails a further claim:

(4)   The fact that a proposition is true is not evidence that believing the proposition would have good or bad ethical or political consequences.

If (3) is true, so is (4). But surely there are counterexamples to proposition (4). Consider a physician who will give a drug to her patients if she thinks the drug is safe but will withhold the drug if she thinks it is not. Suppose that the drug will provide significant health benefits if it is safe. And suppose further that the physician is a pretty good judge of whether the drug is safe. We then have a causal chain in which earlier links raise the probability of later ones:

the drug is safe → the doctor thinks the drug is safe → the patients receive the drug → good consequences accrue to the patients

In this instance, correlation is *transitive*. The nonethical statement "the drug is safe" is therefore evidence for the ethical statement "good consequences accrue to the patients." An ethical and a nonethical fact are evidentially related, just like smoke and fire. I conclude that (4) and therefore (3) are false (Stephens 2000).

It might be replied that (3) and (4) can be saved from refutation by focusing exclusively on the two propositions under discussion. Consider just the two statements "the drug is safe" and "good consequences accrue to the patients." In the absence of any further information, there is no saying whether these statements are positively evidentially relevant to each other, negatively relevant, or entirely irrelevant. We assumed in our story that the physician is well meaning and discerning. This assumption was enough to bring the two statements into a positive relation. However, if the doctor were malevolent or a very bad judge of drug safety, the two statements would stand in a relation of negative relevance or be mutually irrelevant. With no further information, the relation of the two statements is indeterminate. This is a probabilistic analog of Duhem's thesis (Sober 1988).

The trouble with this reply is that what is true for the pair of statements about the physician is true for practically any pair of statements.[2] The evidence relation isn't binary; it has at least three places. When one statement is evidence for a second, this usually is due to the mediation of a third, which provides background information.[3] Given this, it is hardly surprising that ethical facts can provide evidence about scientific propositions; they can do this if one's background assumptions include other, ethical claims.

How is this criticism of (3) and (4) related to Hume's famous dictum that an *ought* cannot be inferred from an *is*? Although Hume was talking about deduction, it is natural to generalize his thesis to a claim about evidential relationships that are nondeductive and probabilistic. What one needs to say here is that an *is*-statement and an *ought*-statement are not evidentially relevant to each other, unless one's background assumptions include other *ought*-statements. But because one's background assumptions often *do* include such "bridge principles," evidential reasoning can run from *is* to *ought* and from *ought* to *is*.[4]

How might propositions (3) and (4) be refined? What is needed is a way to separate the pattern exemplified by the physician from a second class of examples, which William James (1897) addressed in his famous discussion of *the will to believe*. James argued that believing in God can provide substantial psychological benefits. Faith in God can help people feel that their lives are meaningful, which may save them from depression and allow them to lead worthwhile lives. It is debatable how general this point is; there are enough happy atheists and depressed theists around to make one suspect that, at least for many people, mental health is independent of theological conviction. But let's restrict our attention to people who fill James's bill. These people would benefit from believing that God exists; I'll go further and say that it is a *good thing* for

	P is true	P is false
S believes P	w	x
S does not believe P	y	z

Table 5.1. Comparing utilities as a function of what is believed and what is the case.

these people to embrace theism and thus save themselves from the slough of despond. However, I still want to claim that the fact that they would benefit from believing in God provides no evidence that God in fact exists. It is examples like this that make proposition (3) sound so plausible. What distinguishes James's theist from the well-meaning physician?

We can separate these cases by considering the two-by-two table above (table 5.1). The entry in a cell represents *utility*—how good or bad the consequences are of being in that situation. In the physician case (where P="the drug is safe"), there are both "vertical" and "horizontal" effects. The well-being of the patients is affected both by whether the drug is safe $(w>x)$[5] and by what the physician believes $(w>y$ and $z>x)$. In the case of James's believer (where P="God exists"), however, there are only vertical effects. As far as the individual's psychological well-being is concerned, the only thing that matters is that he or she believes in God $(w>y$ and $x>z)$; whether God actually exists doesn't matter $(w=x$ and $y=z)$.

This, I think, provides the key to revising propositions (3) and (4). Our question—when do the ethical consequences of believing a proposition provide evidence as to whether the proposition is true?—can be represented by using the tools of decision theory. We begin by identifying the expected value of each of the two "acts."[6]

$$\text{EV[Believe P]} = wp + x(1-p)$$
$$\text{EV[Don't Believe P]} = yp + z(1-p).$$

Here p denotes the probability that the proposition P is true (and I assume that acts are independent of states of the world). When does the fact that EV[Believe P] > EV[Don't Believe P] provide information about the value of p? A little algebra reveals that

EV[Believe P] > EV[Don't believe P] if and only if $p/(1-p) > (z-x)/(w-y)$.

Suppose that the left-hand inequality is true. When will filling in the values for the utilities w, x, y, and z provide a nontrivial lower or upper bound on the value of p? This fails to happen in the case of James's theist because $(w-y)$ is positive while $(z-x)$ is negative. With these values, all that follows is that $p/(1-p)$ must be greater than some negative number; this is entirely uninformative, since no ratio of probabilities can be negative. The case of the physician is different. Here $(z-x)$ and $(w-y)$ are both positive; therefore, their ratio provides a nontrivial lower bound for the value of p.[7] Thus, science and ethics are not always as separate as propositions (3) and (4) suggest; sometimes information about the ethical consequences of believing a proposition *does* provide information about the probability that the proposition is true.

I have represented the ethical consequences of believing a proposition in terms of the expected utility of doing so. This may sound as if it requires a consequentialist ethics, but in fact it does not. A nonconsequentialist also can enter payoffs in the four cells. Furthermore, the argument is general. Utilities may be calculated in accordance with an ethical theory or on some nonethical basis. What we have here is a general format in which judgments about which acts are better, plus information about the utilities of outcomes, can have implications about the probabilities of propositions.[8]

Before phoning the National Science Foundation with the news that ethics can be a source of evidence for scientific claims, we should reflect on the fact that the expected utility of an action is a composite quantity, built up from probabilities and utilities. If one's ethical judgments about which actions one should perform are based on comparing their expected values, then ethical judgments *require* information about probabilities for their formulation. If so, how can ethical judgments be a source of information about probabilities? This point does not rescue propositions (3) and (4) from counterexample, but it does suggest a new way to think about the problem.

There is a special circumstance in which it is possible to decide which of the two actions (believe P, or don't) is better without any information about the probability that P is true. This is the case in which one action *dominates* the other. James's argument is of this type: He contends that belief in God is beneficial, whether or not God in fact exists.[9] However, we have already seen that the inequality of the expected utilities provides no information about the probabilities in James's case. Without dominance, no conclusion about which action is better can be reached unless one *already* has information about the probabilities. What we have here is an instance of the maxim "out of nothing, nothing comes." If comparing the ethical consequences of believing P and of not believing P has implications about the probability of P, this must be because the description of the ethical consequences already has built into it some information about those probabilities. Thus, ethical facts (about expected utilities) and scientific facts (about probabilities) *are* connected, contrary to what (3) and (4) assert. However, the problem is that this connection is *useless*; we can't use ethical information to gain information we don't already have about probabilities.

The situation would be different if we were able to discover which actions are better than which others, when dominance fails, without already having to have information about probabilities. For example, if there were an infallible guru who would simply tell us what to do, and who would reveal the utilities that go with different states of the world, we could use these inputs to obtain new information about probabilities. But in the absence of such an authority, we are left with the conclusion that our access to information about what we should do must be based on information about probabilities (except when there is dominance). When an ethical conclusion requires information about probabilities (as in the case of the physician), that conclusion can't be a source of new information about those probabilities. And when the ethical conclusion can be reached without information about probabilities (as in the case that James describes), the conclusion tells us nothing about the probabilities. This suggests the following dilemma argument:

> The question of whether believing P has better ethical consequences than not believing P either depends for its answer on information about the probability of P, or it does not.

If it does so depend, then we need information about probabilities to answer the ethical question, and so the ethical judgment cannot supply information about probabilities that we don't already have.

If it does not so depend, then the ethical judgment has no implications about the probability of P.

(5)   Judgments about the ethical consequences of believing P cannot supply *new* information about the probability of P.

The conclusion of this argument, proposition (5), is a reasonable successor to the failed propositions (3) and (4).

The argument just presented is reminiscent of Rudner's (1953) well-known argument that the scientist qua scientist makes value judgments.[10] Rudner describes a physician who must decide whether a drug is safe and argues that this decision must be based on considering ethical features of the four possible outcomes depicted in table 5.1, which we have already discussed. Rudner's argument elicited two criticisms. Levi (1967) contended that accepting a proposition and acting on one's belief are distinct and that the former should not be based on ethical values; Jeffrey (1956) maintained that science is not in the business of accepting and rejecting but merely seeks to assign probabilities to hypotheses. My own argument is neutral on Rudner's position. Perhaps deciding what to believe depends on ethical values; perhaps it does not. My point is that no matter how one decides what to believe, one still can consider what the ethical consequences are of that decision. My question was whether this ethical consideration has implications concerning the probabilities of hypotheses. It does in the case of the physician but not in the case of James's theist.

In the example about the physician, and in many other examples of moral deliberation about which action to perform, one's ethical decision depends on matters of scientific fact. In terms of table 5.1, the ubiquitous pattern is that an inequality gets reversed as one moves from the first column to the second. Although one's ethical decision thus depends on a judgment about a matter of scientific fact, it is possible to form a judgment about the scientific facts without having a commitment, one way or the other, on the ethical question. The physician can't decide whether to administer the drug without knowing something about its probability of being safe, but it is perfectly possible to discover whether a drug is safe without having a view, one way or the other, on whether unsafe drugs should be withheld from patients. Moral ignoramuses can assess the weight of evidence, but scientific ignoramuses cannot make good moral decisions (when those decisions depend, as they almost always do, on scientific matters of fact).

Let's review our progress from propositions (1) and (2) through (3) and (4) and then to (5). Proposition (1) concerns the *behavior of scientists*, whereas (3) and (4) concern the logic of various scientific *concepts*. Proposition (2) ambiguously straddles this distinction; "scientific inference" can be taken to refer to what scientists do or to the formal properties of various types of argument. Proposition (5) addresses the concept of evidence; it does not assert that scientists are immune from political and ethical influence when they decide whether one proposition is evidence for another. More specifically, the claim is not that ethical values (represented by the expected utilities of actions and the utilities of outcomes) *have no implications* about the probabilities of hypotheses, but that ethical inputs are *not needed* to estimate those probabilities. This is why looking to ethics for evidence concerning the truth of scientific hypotheses is to place the cart before the horse.

NOTES

1. I am grateful to Ellery Eells and Dan Hausman for useful comments.

2. The exception arises when one statement deductively entails the other; then they must be positively relevant or of zero relevance, and negative relevance is ruled out.

3. I say "at least" three places because in most scientific contexts, what one can discuss is whether the evidence discriminates between a pair of hypotheses, given a set of background assumptions. See Sober (1994a) for discussion.

4. Does the *ought implies can* principle undermine Hume's thesis? The principle's contrapositive asserts that if it is impossible for an agent to perform an action, then it is false that the agent ought to do so. Here an *is* implies the negation of an *ought*. Hume's thesis can be preserved by insisting that the negation of an *ought*-statement is not itself an *ought*-statement. Similar reasoning is required if one wishes to reconcile Hume's thesis with the fact that philosophers have presented philosophical arguments (of varying quality) for the claim that normative ethical statements lack truth values. These arguments do not contain premises that are normative ethical statements. For example, Harman (1977) and Ruse and Wilson (1986) each present parsimony arguments for the nonexistence of ethical facts; see Sober (2005) and (1994b), respectively, for discussion of each.

5. I take it that it doesn't matter whether the drug is safe if the doctor doesn't believe that it is ($y = z$), because patients won't receive the drug in that situation regardless of whether the drug would be good for them.

6. It might be suggested that believing a proposition is not an action, in the sense that it is not subject to the will. This point is sometimes used against Pascal's wager, but it is an objection that Pascal successfully addressed: He says that if absorbing his argument does not instantly trigger belief, one should go live among religious people so that habits of belief will gradually take hold. Believing a proposition is like other "nonbasic actions": Being president of the United States isn't something one can directly bring about by an act of will, but this does not place it outside the domain of decision theory. For further discussion, see Mougin and Sober (1994).

7. Symmetrically, if not believing P were the better action, this would impose a nontrivial upper bound on the value of the probability.

8. Decision theory from early on has taken an interest in describing the interrelationships of expected utility, utility, and probability. For a brief introduction, see Skyrms (2000, pp. 138–43).

9. Pascal's wager, when the payoffs are finite, does *not* have this property. One needs some information about the probability of God's existing to reach a decision about whether believing is better than not believing. In fact, the theist contemplating Pascal's wager (with finite payoffs) is in the same qualitative situation as the physician deciding whether to believe that the drug is safe. See Mougin and Sober (1994) discussion.

10. The problem that Rudner addresses is the one that James (1897) and W. K. Clifford (1879) debated. It also was central to the debate between the "left" and "right" wings of the Vienna Circle. Neurath argued that evidence does not determine theory choice and that ethical and political values can and should be used to close the gap; Schlick, Carnap, and Reichenbach countered that the intrusion of such values into theory choice is both undesirable and unnecessary: It would compromise the objectivity of science, and scientific inferences can be drawn without taking such values into account. See Howard (2002) for discussion.

REFERENCES

Alvarez, W., and F. Asaro. 1990. "What Caused the Mass Extinction? An Extraterrestrial Impact." *Scientific American*, 263, pp. 78–84.

Clifford, W. K. 1879. "The Ethics of Belief," in *Lectures and Essays*, vol. 2, pp. 177–211. London: Macmillan.

Harman, G. 1977. *The Nature of Morality*. New York: Oxford University Press.

Howard, D. 2002. "Philosophy of Science and Social Responsibility—Some Historical Reflections," in A. Richardson and G. Hardcastle, eds., *Logical Empiricism in North America*. Minneapolis: University of Minnesota Press.

James, W. 1897. "The Will to Believe," in *The Will to Believe and Other Essays in Popular Philosophy*. New York: Longmans Green.

Jeffrey, R. 1956. "Valuation and Acceptance of Scientific Hypotheses." *Philosophy of Science*, 33, pp. 237–46.

Levi, I. 1967. *Gambling with Truth*. Cambridge, MA: MIT Press.

Mougin, G., and E. Sober. 1994. "Betting against Pascal's Wager." *Nous*, 28, pp. 382–95.

Rudner, R. 1953. "The Scientist *Qua* Scientist Makes Value Judgments." *Philosophy of Science*, 20, 1–6.

Ruse, M., and E. Wilson. 1986. "Moral Philosophy as Applied Science." *Philosophy*, 61, pp. 173–92. Reprinted in E. Sober, ed. 1993. *Conceptual Issues in Evolutionary Biology*, 2nd ed. Cambridge, MA: MIT Press.

Skyrms, B. 2000. *Choice and Chance*, 4th ed. Belmont, CA: Wadsworth.

Sober, E. 1988. *Reconstructing the Past—Parsimony, Evolution, and Inference*. Cambridge, MA: MIT Press.

Sober, E. 1994a. "Contrastive Empiricism," in *From a Biological Point of View*. New York: Cambridge University Press.

Sober, E. 1994b. "Prospects for Evolutionary Ethics," in *From a Biological Point of View*. New York: Cambridge University Press.

Sober, E. 2005. *Core Questions in Philosophy*. Upper Saddle River, NJ: Prentice-Hall.

Stephens, C. 2000. "Why Be Rational? Prudence, Rational Belief and Evolution." PhD dissertation, University of Wisconsin, Madison.

SIX

REJECTING THE IDEAL OF
VALUE–FREE SCIENCE

Heather Douglas

6.1 Introduction

The debate over whether science should be value free has shifted its ground in the past sixty years. As a way to hold science above the brutal cultural differences apparent in the 1930s and 1940s, philosophers posited the context of discovery–context of justification distinction, preserving the context of justification for reason and evidence alone. It was in the context of justification that science remained free from subjective and/or cultural idiosyncrasies; it was in the context of justification that science could base its pursuit of truth. Even within the context of justification, however, values could not be completely excluded. Several philosophers in the 1950s and 1960s noted that scientists needed additional guidance for theory choice beyond just logic and evidence alone. (See, for example, Churchman 1956, Levi 1962, or Kuhn 1977.) *Epistemic values* became a term to encompass the values acceptable in science as guidance for theory choice. Some argued that only these values could legitimately be part of scientific reasoning or that it was the long-term goal to eliminate nonepistemic values (McMullin 1983). By 1980, "value-free science" really meant science free of nonepistemic values.

But not all aspects of science were to hold to this norm. As the distinction between discovery and justification has been replaced by a more thorough account of the scientific process, the limits of the "value-free" turf in science have become clearer. It has been widely acknowledged that science requires the use of nonepistemic values in the "external" parts of science, that is, the choice of projects, limitations of methodology (particularly with respect to the use of human subjects), and the application of science-related technologies.[1] So the term *value-free science* really refers to the norm of epistemic values only in the internal stages of science. It is this qualified form of "value-free science" that is held up as an ideal for science.

Many assaults can and have been made on this ideal. It has been argued that it is simply not attainable. It has been argued that the distinction between epistemic and nonepistemic is not clear enough to support the normative weight placed on the distinction by the ideal. (I have argued this elsewhere [Machamer and Douglas 1999], as have Rooney 1992 and Longino 1996, more eloquently.) One can, however, take a stronger tack than the claim that value-free science is an unattainable or untenable ideal. One can argue that the ideal itself is simply a bad ideal. As I have argued in greater detail elsewhere, in many areas of science, particularly areas used to inform public policy decisions, science should not be value free, in the sense just described (Douglas 2000). In these areas of science, value-free science is neither an ideal nor an illusion. It is unacceptable science.

Rejecting the ideal of value-free science, however, disturbs many in the philosophy of science. The belief persists that if we accept the presence of values (particularly nonepistemic values) in the inner working of science, we will destroy science and set ourselves adrift on the restless seas of relativism. At the very least, it would be a fatal blow to objectivity. As Hugh Lacey has recently warned, without the value-free ideal for science's internal reasoning, we would lose "all prospects of gaining significant knowledge" (Lacey 1999, 216).

I disagree with this pessimistic prediction and instead think that rejecting the value-free ideal would be good for science by allowing for more open discussion of the factors that enter into scientific judgments and the experimental process. In this chapter, I will first explain why nonepistemic values are logically needed for reasoning in science, even in the internal stages of the process. I will then bolster the point with an examination of ways to block this necessity, all of which prove unsatisfactory. Finally, I will argue that rejection of the value-free ideal does not demolish science's objectivity and that we have plenty of remaining resources with which to understand and evaluate the objectivity of science. By understanding science as value laden, we can better understand the nature of scientific controversy in many cases and even help speed resolution of those controversies.

6.2 Choices and Values in Science

To make the normative argument that values are required for good reasoning in science, I will first describe the way in which values play a crucial decision-making role in science, which I will then briefly illustrate. The areas of science with which I am concerned are those that have clear uses for decision making. I am not focused here on science used to develop new technologies, which then are applied in various contexts. Instead, I am interested in science that is used to make decisions, science that is applied as useful knowledge to select courses of action, particularly in public policy.

One hundred years ago, science was little used in shaping public policy. Indeed, the bureaucracies that now routinely rely of scientific expertise in their decision making were either nonexistent in the United States (e.g., Environmental Protection Agency, Consumer Product Safety Commission, Occupational Safety and Health Administration, Department of Energy) or in their earliest stages of development (Food and Drug Administration, Centers for Disease Control). Now, entire journals (*Chemosphere, Journal of Applied Toxicology and Pharmacology, CDC Update,* etc.), institutions (e.g., National Institute for Environmental Health Sciences, Chemical Industry Institute of Toxicology, National Research Council), and careers are devoted to science that will be used to develop public policy. Although science is used to make decisions in other spheres as well (e.g., in the corporate world and nongovernmental organizations), I will draw my examples from the use of science in public policy. It is in this realm that the importance of scientific input is the clearest, with the starkest implications for our views on science.

In the doing of science, whether for use or for pure curiosity, scientists must make choices. They choose a particular methodological approach. They make decisions on how to characterize events for recording as data. They decide how to interpret their results.[2] Scientific papers are usually structured along these lines, with three internal sections packaged within an introduction and a concluding discussion. In the internal sections of the paper (methodology, data, results), scientists rarely explicitly discuss the choices that they make. Instead, they describe what they did, with no mention of alternative paths they might have taken.[3] To discuss the choices that they make would require some justification for those choices, and this is territory the scientist would prefer to avoid. It is precisely in these choices that values, both epistemic and, more controversially, nonepistemic, play a crucial role. Because scientists do not recognize a legitimate role for values in science (it would damage "objectivity"), scientists avoid discussion of the choices they make.

How do the choices require the consideration of epistemic and nonepistemic values? Any choice involves the possibility for error. One may select a methodological approach that is not as sensitive or appropriate for the area of concern as one thinks it is, leading to inaccurate results. One may incorrectly characterize one's data. One may rely upon inaccurate background assumptions in the interpretation of one's results.[4] When the science is used to make public policy decisions, such errors lead to clear nonepistemic consequences. If one is to weigh which errors are more serious, one will need to assign values to the various likely consequences. Only with such evaluations of likely error consequences can one decide whether, given the uncertainty and the importance of avoiding particular errors, a decision is truly appropriate. Thus values become an important, although not determining, factor in making internal scientific choices.

Clearly, there are cases where such value considerations will play a minor or even nonexistent role. For example, there may be cases where the uncertainty is so small that the scientists have to stretch their imaginations to create any uncertainty at all. Or there may be cases where the consequences of error are completely opaque and we could not expect anyone to clearly foresee them. However, I contend that in many cases, there are fairly clear consequences of error (as there are fairly well-recognized practices for how science is used to make policy) and that there is significant uncertainty, generating heated debate among scientists.

In general, if there is widely recognized uncertainty and thus a significant chance of error,[3] we hold people responsible for considering the consequences of error as part of their decision-making process. Although the error rates may be the same in two contexts, if the consequences of error are serious in one case and trivial in the other, we expect decisions to be different. Thus the emergency room avoids as much as possible any false negatives with respect to potential heart attack victims and accepts a very high rate of false positives in the process. (A false negative occurs when one rejects the hypothesis—in this case, that someone is having a heart attack—when the hypothesis is true. A false positive occurs when one accepts the hypothesis as true when it is false.) In contrast, the justice system attempts to avoid false positives, accepting some rate of false negatives in the process. Even in less institutional settings, we expect people to consider the consequences of error, hence the existence of reckless endangerment and reckless driving charges. We might decide to isolate scientists from having to think about the consequences of their errors. I will discuss this line of thought later. But for now, let us suppose that we want to hold scientists to the same standards as everyone else and thus that scientists *should* think about the potential consequences of error.

In science relevant to public policy, the consequences of error clearly include nonepistemic consequences. Even the most internal aspects of scientific practice—the characterization of events as data—can include significant uncertainty and clear nonepistemic consequences of error. An example I have discussed elsewhere that effectively demonstrates this point is the characterization of rat liver tissue from rats exposed to dioxin. (See Douglas 2000 for a more complete discussion.) In a key study used for setting regulatory policy completed in 1978, four groups of rats were exposed to three different dose levels of dioxin (2,3,7,8-tetrachloro-dibenzo-p-dioxin) plus a control group (Kociba et al. 1978). After two years of dosing, the rats were killed and autopsied. Particular focus was placed on the livers of the rats, and slides were made of the rat liver tissues, which were then characterized as containing tumors, benign or malignant, or being free from such changes. Over the next fourteen years, those slides were reevaluated by three different groups, producing different conclusions about the liver cancer rates in those rats. Clearly, there is uncertainty about what should count as liver cancer in rats and what should not.

What does this uncertainty mean for the decision of whether to characterize or not characterize a tissue slide as containing a cancerous lesion? In an area with this much uncertainty, the scientist risks false positives and false negatives with each characterization. Which errors should be more carefully avoided? Too many false negatives are likely to make dioxin appear to be a less potent carcinogen, leading to weaker regulations. This is precisely what resulted from the 1990s industry-sponsored reevaluation (see Brown 1991) that was used to weaken Maine water-quality standards. Too many false positives, on the other hand, are likely to make dioxin appear to be more potent and dangerous, leading to burdensome and unnecessary overregulation. Which consequence is worse? Which error should be more scrupulously avoided? Answering these questions requires reflection on ethical and societal values concerning human health and economic vitality. Such reflection is needed for those uncertain judgments *at the heart* of doing science.

One might counter this line of thought with the suggestion that scientists not actually make the uncertain judgments needed to proceed with science but, instead, that scientists estimate the uncertainty in any given judgment and then propagate that uncertainty through the experiment and analysis, incorporating it into the final result.[6] Two problems confront this line of thought. The first is purely practical. If the choices scientists must make occur early in the process, for example, a key methodological choice, it can be quite difficult to estimate precisely the effect of that choice on the experiment. Without a precise estimate, the impact on the experiment cannot be propagated through the experimental analysis. For example, in epidemiological studies, scientists often rely on death certificates to determine the cause of death of their subjects. Death certificates are known to be wrong on occasion, however, and to be particularly unreliable for some diseases, such as soft-tissue sarcoma (Suruda, Ward, and Fingerhut 1993) The error rates for rare diseases like soft-tissue sarcoma are not well known, however, and other sources of data for epidemiological studies are difficult or very expensive to come by. Expecting scientists to propagate a precise estimate of uncertainty about their source of data, in this case, through a study, would be unreasonable.

The second problem is more fundamental. To propagate the uncertainty, the scientist must first estimate the uncertainty, usually making a probabilistic estimate of the chance of error. But how reliable is that estimate? What is the chance of error in the estimate, and is the chance low enough to be acceptable? Making this kind of judgment again must involve values to determine what would be acceptable. Having scientists make estimates of uncertainty pushes the value judgments back one level but does not eliminate them. (This problem is first discussed in Rudner 1953 and, to my knowledge, is not addressed by his critics.) The attempt to escape the need for value judgments with error estimates merely creates a regress, pushing back the point of judgment further from view and making open discussion about the judgments all the more difficult. This serves to obscure the important choices and values involved, but it does not eliminate them.

Thus, *if* we want to hold scientists to the same responsibilities the rest of us have, the judgments needed to do science cannot escape the consideration of potential consequences, both intended and unintended, both epistemically relevant and socially relevant. This is not to say that evidence and values are the same thing. Clearly, logically, they are not. Values are statements of norms, goals, and desires; evidence consists of descriptive statements about the world. Hume's prohibition remains in effect; one cannot derive an *ought* from an *is*. This does not mean, however, that a descriptive statement is free from values in its origins. Value judgments are needed to determine whether a descriptive label is accurate enough and whether the errors that could arise from the description call for more careful accounts or a shift in descriptive language. Evidence and values are different things, but they become inextricably intermixed in our accounts of the world.

6.3 Scientists, Responsibility, and Autonomy

Although I hope to have convinced my reader by now that nonepistemic values do have a legitimate role to play in science and are needed for good reasoning, one still may wish to shield scientists from having to make value judgments as part of their work. There are two general and related objections to my position that can be made: (1) Scientists shouldn't make choices involving value judgments—they should do their science concerned with epistemic values only and leave determining the implications of that work to the policy makers, and (2) we should shield scientists from having to think about the consequences of error in their work in order to protect the "value neutrality" of the scientific process. I will address each of these in turn.

When the issue of values in science was raised in the 1950s by Churchman and Rudner, the response to their suggestion that values played an important role in science was that scientists do not need to consider values because they are not the ones performing the decisions for which consequences of error are relevant and/or they are simply reporting their data for the use of decision makers. The example of rat liver characterization choices from the previous section demonstrates the difficulty of holding to a "reporting data only" view of scientists' role in public policy. Even in the act of reporting "raw" data, some decisions are made as to how to characterize events in turning those events into raw data. (I also argued previously that reporting raw data with uncertainty estimates does not free the statements from relying in part on value judgments.) Those choices involve the potential for error and, in the example, clear and predictable consequences of error. Thus, even raw data can include judgments of characterization that require values in the process.

Scientists, however, rarely report solely raw data to public decision makers. They are usually also called on to interpret that data, and this is to the good. It would be a disaster for good decision making if those with far less expertise than climatologists, for example, were left with the task of interpreting world temperature data. Policy makers rarely have the requisite expertise to interpret data, and it is fitting that scientists are called on to make some sense of their data. Yet scientists' selection of interpretations involves selection of background assumptions, among other things, with which to interpret the data.

For example, in toxicology, there is a broad debate about whether it is reasonable to assume that thresholds exist for certain classes of carcinogens, or whether some other function (e.g., some extrapolation toward zero dose and zero response) better describes their dose-response relationship. There are complex sets of background assumptions supporting several different interpretations of dose-response data sets, including assumptions about the biochemical mechanisms at work in any particular case. Which background assumptions should be selected? Depending on which background assumptions one adopts, the threshold model looks more or less appropriate. In making the selection of background assumptions, not only epistemic considerations should be used but also nonepistemic considerations, such as which kinds of errors are more likely, given different sets, and how we weigh the seriousness of those errors. In short, we cannot effectively use scientific information without scientific interpretation, but interpretation involves value considerations. And because few outside the scientific community are equipped to make those interpretations, scientists usually must interpret their findings for policy makers and the public.

Still, to preserve the value-free ideal for useful science, one might be tempted to argue that we need to insulate scientists from considering the consequences of scientific error (the second objection). Perhaps we should set scientists apart from the general moral requirements to which most of us are held. Perhaps scientists should be required to search solely for truth, and any errors they make along the way (and the consequences of those errors) should be accepted as the cost of truth by the rest of society. Under this view, scientists may make dubious choices with severe consequences of error, but we would not ask them to think about those consequences and would not hold them responsible if and when they occur.

In considering this line of thought, it must be noted that, in other areas of modern life, we are required to consider unintended consequences of actions and to weigh benefits against risks; if we fail to do so properly, we are considering negligent or reckless. Scientists can be held exempt from such general requirements only if (1) we thought that epistemic values always trumped social values *or* (2) someone else could take up the burden of oversight. If we thought that epistemic values were a supreme good, they would outweigh social and moral values every time, and thus scientists would not need to consider nonepistemic values. If, on the other hand, someone else (with the authority to make decisions regarding research choices) were set up to consider nonepistemic values and social consequences, scientists could be free of the burden. If both of these options fail, the burden of responsibility to consider *all* the relevant potential consequences of one's choices falls back to the scientist. Let me consider each of these possibilities in turn.[7]

Do epistemic values trump other kinds of values? Is the search for truth (or knowledge) held in such high esteem that all other values are irrelevant before it? If we thought the search for truth (however defined, and even if never attained) was a value in a class by itself, worth all sacrifices, then epistemic values alone would be sufficient for considering the consequences of research. Epistemic values would trump all other values, and there would be no need to weigh them against other values. However, there is substantial evidence that we do not accord epistemic values such a high status. That we place limits on the use of human (and now animal) subjects for their use in research indicates we are not willing to sacrifice all for the search for truth. That our society has struggled to define an appropriate budget for federally funded research, and that some high-profile projects (such as the Mohole project in the 1960s[8] and the superconducting supercollider project in the 1990s) have been cut altogether suggests that in fact we do weigh epistemic values and goals against other considerations. That epistemic values are important to our society is laudable, but so, too, is that they are not held transcendently important when compared with social or ethical values. The first option to escape the burden of nonepistemic reflection is closed to scientists.

The second option remains but is fraught with difficulties. We could acknowledge the need to reflect on both social and epistemic considerations (i.e., the intended outcomes, the potential for errors and their consequences, and the values needed to weigh those outcomes) but suggest that someone besides scientists do the considering. We may find this alternative attractive because we have been disappointed by scientists' judgments in the past (and the values that shaped those judgments) or because we want to maintain the purity of science, free from social values.[9] The costs of nonepistemic research oversight by outsiders, however, outweigh the potential benefits.

For this option to be viable, consideration of nonepistemic consequences cannot be an afterthought to the research project; instead, it must be an integral part of it.[10] Those shouldering the full social and ethical responsibilities of scientists would have to have decision-making authority with the scientists, in the same way that research review boards now have the authority to shape methodological approaches of scientists when they are dealing with human subjects. However, unlike these review boards, whose review takes place at one stage in the research project, those considering nonepistemic consequences of scientific choices would have to be kept abreast with the research program at every stage (where choices are being made) and would have to have the authority to change those choices if necessary. Otherwise, the responsibility would be toothless and thus meaningless.

To set up such a system would be to dilute any decision-making autonomy the scientists have between the scientists and their ethical overseers. This division of authority would probably lead to resentment among the scientists and to reduced reflection by scientists on the potential consequences of research. After all, increased reflection would only complicate the scientist's research by requiring more intensive consultation with the ethical overseer. Without the scientists' cooperation in considering potential consequences, the overseers attempting to shoulder the responsibility for thinking about the consequences of science and error would be blind to some of the more important ones.

To see why, consider that scientists performing the research may in many cases be the only ones who are both aware of the uncertainties and potential for error and of the likely or foreseeable consequences of error. For example, before the Trinity test in 1945, several theoretical physicists realized there was a possibility a nuclear explosion might ignite the atmosphere. Hans Bethe explored this possibility and determined that the probability was infinitesimally small. Who else could have thought of this potential for error and followed it through sufficiently to determine that the chance of this error was sufficiently small to be disregarded? This is a dramatic example, but it serves to illustrate that we need scientists to consider where error might occur and what its effects might be. Few outside Los Alamos could have conceived of this possibility, much less determined it was so unlikely that it was not a worry. Only with the active reflection of scientists on the edge of the unknown can the responsibilities be properly met.

Thus, the responsibility to consider the social and ethical consequences of one's actions and potential error cannot be sloughed off by scientists to someone else without a severe loss of autonomy in research. We have no adequate justification for allowing scientists to maintain nonepistemic blinders on an ongoing basis. Because both epistemic and nonepistemic values are important, scientists must consider both when making choices with consequences relevant to both. To keep scientists from considering the consequences of their work would be a highly dangerous approach (for science and society), with risks far outweighing any benefits. However, some might still insist that the damage to the objectivity of science caused by accepting a legitimate role for nonepistemic values in scientific reasoning would be so severe that we should still attempt to shield scientists (somehow) from that responsibility. I will argue in the next section that objectivity is robust enough without needing to be defined in terms of the value-free ideal.

6.4 Implications for Objectivity and Science

Objectivity is one of the most frequently invoked yet vaguely defined concepts in the philosophy of science.[11] Happily, in recent years, some nuanced philosophical and historical work has been done to attempt to clarify this crucial and vague term. What has become apparent in most of this work is that objectivity is an umbrella concept encompassing a broad, interrelated, but irreducibly complex set of meanings. For example, in the philosophical literature of the past decade, several authors have pointed out that objectivity has, in fact, multiple meanings already in play (see, e.g., Lloyd 1995; Fine 1998). Historical work has suggested how this could come about, with detailed work tracking how the meaning of *objectivity* has shifted and accrued new nuances over the past three centuries (Daston and Gallison 1992; Daston 1992; Porter 1992, 1995). I will argue in this section that we can discard the value-free meaning of objectivity without significant damage to the concept overall. Despite the long association between "value free" and "objective," there is nothing necessary about the link between the two concepts.

Before embarking on a description of objectivity's complexity, I should make clear that not all of the other traditional meanings associated with objectivity are discussed here. Some of the meanings attached to objectivity are functionally unhelpful for evaluating whether a statement, claim, or outcome is, in fact, objective. For evaluating the objectivity of science, we need operationalizable definitions, definitions that can be applied to deciding whether something is actually objective. This restriction eliminates from consideration some of the more metaphysical notions of objectivity, such as an aperspectival perspective or being independent of human thought. Because we currently have no way of getting at these notions of objectivity, they are unhelpful for evaluating the objectivity of science or the objectivity of other human endeavors. I will not consider them here.

Even without functionally useless aspects of objectivity, there are seven distinct meanings for objectivity, aside from "value free"; that is, there are seven clear and accessible ways that we can mean "objective" without meaning "value-free." This result suggests that there are considerable resources inherent in the term *objectivity* for handling the rejection of the value-free ideal. Let me elaborate on these seven alternatives.[12]

Two of the senses of *objectivity* apply to situations where we are looking at human interactions with the world. The first is perhaps the most powerfully persuasive at convincing ourselves we have gotten ahold of some aspect of the world: manipulable objectivity. This sense of objectivity can be invoked when we have sufficiently gotten at the objects of interest such that we can use those objects to intervene reliably elsewhere. As with Ian Hacking's famous example from *Representing and Intervening*, scientists don't doubt the objective existence of electrons when they can use them to reliably produce images of entirely different things with an electron-scanning microscope (Hacking 1983, 263). Our confidence in the objective existence of the electron should not extend to all theoretical aspects connected to the entity—the theory about it may be wrong, or the entity may prove to be more than one thing—but it is difficult to doubt that some aspect of the world is really *there* when one can manipulate it as a tool consistently.

In cases where some scientific entity can be used to intervene in the world and that intervention can be clearly demonstrated to be successful, we have little doubt about the manipulable objectivity (sense 1) of the science. However, the controversial cases of science and policy today do not allow for a clear check on this sense of objectivity. The science in these cases concerns complex causal systems that are fully represented only in the real world, and to attempt to do the intervention tests in the real world would be unethical or on such long time scales as to be useless (or both). Imagine, for example, deliberately manipulating the global climate for experimental purposes. Not only would the tests take decades, not only would it expose world populations to risks from climate change, but also it still would not be conclusive; factors such as variability in sun intensity and the length of time needed to equilibrate global carbon cycles make intervention tests hugely impractical. It is very doubtful that we will have a sense of manipulable objectivity for cases such as these.

For some of these cases, there is another potentially applicable meaning for objectivity, one that trades on multiple avenues of approach.

If we can approach an object through different and hopefully independent methods and if the same object continues to appear, we have increasing confidence in the object's existence. The sense of objectivity
invoked here, convergent objectivity (sense 2), is commonly relied on in
scientific fields where intervention is not possible or ethical, such as as-
tronomy, evolutionary biology, and global climate studies.[13] When evidence from disparate areas of research points toward the same result or
when epistemically independent methodologies produce the same answer, our confidence in the objectivity (in this sense) of the result increases. (See Kosso 1989 for a discussion of the problem of epistemic independence.) We still might be fooled by an objectively convergent result;
the methods may not really be independent, or some random conver-
gence may be occurring. But objectivity is no *guarantee* of accuracy; instead, it is the best we can do.

In addition to these two senses of objectivity focused on human interactions with the world, there are senses of objectivity that focus on individual thought processes. It is in this category that one would place
the "value-free" meaning of objectivity. As I argued previously, this sense
of *objective* should be rejected as an ideal in science. It can be replaced
with two other possibilities: detached objectivity or value-neutral objectivity. *Detached objectivity* refers to the prohibition against using values
in place of evidence. Simply because one wants something to be true
does not make it so, and one's values should not blind one to the existence of unpleasant evidence. Now it may seem that my defense of detached objectivity contradicts my rejection of value-free objectivity, but
closer examination of the role of values in the reasoning process shows
that this is not the case. In my preceding discussion and examples, values neither supplant nor become evidence by themselves; they do shape
what one makes of the available evidence. One can (and should) use
values to determine how heavy a burden of proof should be placed on a
claim and which errors are more tolerable. Because of the need for
judgments in science throughout the research process, values have legitimate roles to play throughout the process. But using values to blind
one to evidence one would rather not see is not one of those legitimate
roles. Values cannot act in place of evidence; they can only help determine how much evidence we require before acceptance of a claim. The
difference between detached objectivity (sense 3) and value-free objectivity is thus a crucial one.

Value-neutral objectivity should also not be confused with value-free objectivity. In value-neutral objectivity (sense 4), a value position that is neutral on the spectrum of debate, a midrange position that takes no strong stance, is used to inform the necessary judgments. Value-neutral objectivity can be helpful when there is legitimate and ongoing debate over which value positions we ought to hold, but some judgments based on some value position are needed for research and decision making to go forward. Value-neutral objectivity has limited applicability, however; it is not desirable in all contexts. For example, if racist or sexist values are on one side of the relevant value spectrum, value neutrality would not be acceptable, because racist and sexist values have been rightly and soundly rejected. We have good moral reasons for not accepting racist or sexist values, and thus other values should not be balanced against them. Many conflicts involving science and society reflect unsettled debates, however, and in these cases, value neutrality, taking a reflectively balanced value position, can be usefully objective.

I have presented four alternative meanings for objectivity in addition to value free. There are three remaining, all concerned with social processes. The possibility of social processes undergirding objectivity has received increased attention recently, and in examining that body of work, I found three distinct senses of objectivity that relate to social processes: procedural objectivity, concordant objectivity, and interactive objectivity. Procedural objectivity (sense 5) occurs when a process is set up such that regardless of who is performing that process, the same outcome is always produced. (This sense is drawn from Megill 1994; Porter 1992, 1995.) One can think of the grading of multiple-choice exams as procedurally objective, or the rigid rules that govern bureaucratic processes. Such rules eliminate the need for personal judgment (or at least aim to), thus producing "objectivity."

Concordant objectivity (sense 6) occurs when a group of people all agree on an outcome, be it a description of an observation or a judgment of an event. The agreement in concordant objectivity, however, is not one achieved by group discussion or by following a rigid process; it simply occurs. When a group of independent observers all agree that something is the case, their agreement bolsters our confidence that their assessment is objective. This intersubjective agreement has been considered by some essential to scientific objectivity; as Quine wrote:

"The requirement of intersubjectivity is what makes science objective" (1992, 5).

Some philosophers of science have come to see this intersubjective component less as a naturally emergent agreement and more as the result of the intense debate that occurs within the scientific community (Longino 1990; Kitcher 1993; Hull 1988). Agreement achieved by intensive discussion I have termed *interactive objectivity* (sense 7). Interactive objectivity occurs when an appropriately constituted group of people meet and discuss what the outcome should be. The difficulty with interactive objectivity lies with the details of this process: What is an appropriately constituted group? How diverse and with what expertise? How are the discussions to be framed? And what counts as agreement reached among the members of the group? Much work needs to be done to fully address these questions. Yet it is precisely these questions that are being dealt with in practice by scientists working with policy-relevant research. Questions of whether peer review panels for science-based regulatory documents are appropriately constituted and what weight to put on minority opinions and questions of whether consensus should be an end goal of such panels and what defines consensus are continually faced by scientists.

I will not attempt to answer these difficult questions here. The point of describing these seven aspects of objectivity is to make clear that value free is not an essential aspect of objectivity. Rather, even when rejecting the ideal of value-free science, we are left with seven remaining aspects of objectivity with which to work. This embarrassment of riches suggests that rejecting the ideal of value-free science is no threat to the objectivity of science. Not all of the remaining aspects of objectivity will be applicable in any given context (they are not all appropriate), but there are enough to draw on that we can find some basis for the trust we place in scientific results.

6.5 Conclusion

Rejecting the ideal of value-free science is thus uncatastrophic for scientific objectivity. It is also required by basic norms of moral responsibility and the reasoning needed to do sound, acceptable science. It does imply increased reflection by scientists on the nonepistemic implications and potential consequences of their work. Being a scientist per se does not exclude one from that burden. Some scientists may object that their work has no implications for society and that there are no potential nonepistemic consequences of error. Does the argument presented here apply to all of science? My argument clearly applies to all areas of science that have an actual impact on human practices. It may not apply to some areas of research conducted for pure curiosity (at present). But it is doubtful that these two "types" of science can be cleanly (or permanently) demarcated from each other. The fact that one can think of examples at either extreme does not mean there is a bright line between these two types (the useful and the useless) or that such a line would be stable over time.[14] In any case, debates over whether there are clear and significant societal consequences of error in particular research areas would be a welcome change from the assertion that nonepistemic values should play no role in science. Understanding science in this way will require a rejoining of science with moral, political, and social values.

I would like to close this chapter by suggesting that opening the discourse of science to include discussion of nonepistemic values relevant to inductive risks will make answering questions about how to conduct good science easier, not harder. If the values that are required to make scientific judgments are made explicit, it will be easier to pinpoint where choices are being made and why scientists disagree with each other in key cases. It will also make it clearer to the science-observing public the importance of debates about what our values should be. Currently, too many hope that science will give us certain answers on what is the case so that it will be clear what we should do. This is a mistake, given the inherent uncertainty in empirical research. If, on the other hand, values can be agreed on, agreement will be easier to reach about how to best make scientific decisions (for example, as we now have clear guidelines and mechanisms for the use of human subjects in research) and about what we should do regarding the difficult public policy issues we face. If values cannot be agreed on, the source and nature of disagreement can be more easily located and more honestly discussed. Giving up on the ideal of value-free science allows a clearer discussion of scientific disagreements that already exist and may lead to a speedier and more transparent resolution of these ongoing disputes.

ACKNOWLEDGMENTS

My thanks to the University of Puget Sound (Martin Nelson Junior Sabbatical Fellowship) and the National Science Foundation (SDEST grant #0115258) for their support during the writing of this chapter. Also thanks to Harold Kincaid, Wayne Riggs, Alison Wylie, and Ted Richards for their detailed and insightful comments on earlier versions of this chapter. I gave one of those versions at the Eastern APA in December 2001, and the vibrant discussion that followed also helped clarify this work. All remaining muddles are mine alone.

NOTES

1. Rudner (1953, 1) noted the importance of values for the selection of problems. Nagel (1961, 485–87) and Hempel (1965, 90) also noted this necessary aspect of values in science. Rescher (1965) provides a more comprehensive account of multiple roles for values in science, as does Longino (1990, 83–102).

2. Richard Rudner (1953) made a similar point about the practice of science, although Rudner focused solely on the scientist's choice of a theory as acceptable or unacceptable, a choice placed at the end of the "internal" scientific process.

3. It can be difficult to pinpoint where scientists make choices when reading their published work. One can determine that choices are being made by reading many different studies within a narrow area and seeing that different studies are performed and interpreted differently. With many cross-study comparisons within a field, the fact that alternatives are available, and thus that choices are being made, becomes apparent.

4. Note that in reading a scientific paper with any one of these kinds of errors, it would not be necessarily obvious that a choice had been made, much less an error.

5. "Significant chance of error" is obviously a vague term, and whether it applies in different cases can be a serious source of debate. The fact that there is no bright line for whether a chance of error is significant does not mean that one need not think about that chance at all.

6. This is distinct from asking scientists to not consider consequences of error at all, to be addressed later.

7. I have argued these points in greater detail in Douglas 2003.

8. See Greenberg 1967, chapter 9, for a detailed account of Mohole's rise and fall.

9. Someone else may need to do some reflective considering in addition to scientists, but that would still leave the presumption of responsibility with the scientists.

10. There may be special cases where we decide to let scientists proceed without considering what might go wrong and whom it might harm, but these cases would have to be specifically decided, given the research context, and then still carefully monitored. What makes science exciting is its discovery of the new and the unknown. It is difficult to be certain at the beginning of a research project that no serious consequences (either of error or of correct results) lurk in the hidden future.

11. In comparison, consider the concept of truth. Although it, too, is often invoked, much effort has been spent trying to precisely define what is meant. With objectivity, in contrast, it is often assumed that we just "know" what we mean.

12. See Douglas 2004 for a more detailed discussion of these aspects of objectivity.

13. One can create controlled laboratory conditions for small-scale climate studies or evolutionary studies, but there is always a debate over whether all of the relevant factors from the global context were adequately captured by the controlled study.

14. The example of nuclear physics is instructive. Once thought to be a completely esoteric and useless area of research, it quite rapidly (between December 1938 and February 1939) came to be recognized as an area of research with immense potential practical implications.

REFERENCES

Brown, W. R. 1991. "Implication of the Reexamination of the Liver Sections from the TCDD Chronic Rat Bioassay," in M. Gallo, R. J. Scheuplein, and K. A. Van der Heijden, eds., *Biological Basis for Risk Assessment of Dioxins and Related Compounds*, pp. 13–26. Cold Spring Harbor, NY: Cold Spring Harbor Laboratory Press.

Churchman, C. W. 1948. "Statistics, Pragmatics, and Induction." *Philosophy of Science*, 15, pp. 249–68.

Churchman, C. W. 1956. "Science and Decision-Making." *Philosophy of Science*, 22, pp. 247–49.

Daston, L. 1992. "Objectivity and the Escape from Perspective." *Social Studies of Science*, 22, pp. 597–618.

Daston, L., and P. Gallison. 1992. "The Image of Objectivity." *Representations*, 40, pp. 81–128.

Douglas, H. 2000. "Inductive Risk and Values in Science." *Philosophy of Science*, 67, 559–79.

Douglas, H. 2003. "The Moral Responsibilities of Scientists: Tensions between Autonomy and Responsibility." *American Philosophical Quarterly*, 40, pp. 59–68.

Douglas, H. 2004. "The Irreducible Complexity of Objectivity." *Synthese*, 138, pp. 453–73.

Fine, A. 1998. "The Viewpoint of No-One in Particular." *Proceedings and Addresses of the APA*, 72, pp. 9–20.

Greenberg, D. S. 1967. *The Politics of Pure Science*. Chicago: University of Chicago Press.

Hacking, I. 1983. *Representing and Intervening*. New York: Cambridge University Press.

Hempel, C. G. 1965. "Science and Human Values," in *Aspects of Scientific Explanation and Other Essays in the Philosophy of Science*, pp. 81–96. New York: Free Press.

Hull, D. 1988. *Science as a Process*. Chicago: University of Chicago Press.

Jeffrey, R. C. 1956. "Valuation and Acceptance of Scientific Hypotheses." *Philosophy of Science*, 22, 237–46.

Kitcher, P. 1993. *The Advancement of Science*. New York: Oxford University Press.

Kociba, R. J., et al. 1978. "Results of a Two-Year Chronic Toxicity and Oncogenicity Study of 2.3.7.8-Tetrachlorodibenzo-p-Dioxin in Rats." *Toxicology and Applied Pharmacology*, 46, pp. 279–303.

Kosso, P. 1989. "Science and Objectivity." *Journal of Philosophy*, 86, pp. 245–57.

Kuhn, T. 1977. "Objectivity, Value, and Theory Choice," in *The Essential Tension*, pp. 320–39. Chicago: University of Chicago Press.

Lacey, H. 1999. *Is Science Value Free?* New York: Routledge.

Levi, I. 1962. "On the Seriousness of Mistakes." *Philosophy of Science*, 29, pp. 47–65.

Lloyd, E. 1995. "Objectivity and the Double Standard for Feminist Epistemologies." *Synthese*, 104, pp. 351–81.

Longino, H. E. 1990. *Science as Social Knowledge: Values and Objectivity in Scientific Inquiry*. Princeton, NJ: Princeton University Press.

Longino, H. E. 1996. "Cognitive and Non-Cognitive Values in Science: Rethinking the Dichotomy," in L. H. Nelson and J. Nelson, eds., *Feminism, Science, and the Philosophy of Science*, pp. 39–58. Dordrecht: Kluwer.

Machamer, P., and H. Douglas. 1999. "Cognitive and Social Values." *Science and Education*, 8, pp. 45–54.

McMullin, E. 1983. "Values in Science," in P. D. Asquith and T. Nickles, eds., *Proceedings of the 1982 Biennial Meeting of the Philosophy of Science Association*, vol. 1, pp. 3–28. East Lansing, MI: Philosophy of Science Association.

Megill, A. 1994. "Introduction: Four Senses of Objectivity," in A. Megill, ed., *Rethinking Objectivity*, pp. 1–20. Durham, NC: Duke University Press.

Nagel, E. 1961. *The Structure of Science: Problems in the Logic of Scientific Explanation*. New York: Harcourt, Brace, and World.

Porter, T. 1992. "Quantification and the Accounting Ideal in Science." *Social Studies of Science*, 22, pp. 633–52.

Porter, T. 1995. *Trust in Numbers: The Pursuit of Objectivity in Science and Public Life*. Princeton, NJ: Princeton University Press.

Quine, W. V. 1992. *The Pursuit of Truth*. Cambridge, MA: Harvard University Press.

Rescher, N. 1965. "The Ethical Dimension of Scientific Research," in Robert Colodny, ed., *Beyond the Edge of Certainty*, pp. 261–76. Englewood Cliffs, NJ: Prentice-Hall.

Rooney, P. 1992. "On Values in Science: Is the Epistemic/Non-Epistemic Distinction Useful?" in D. Hull, M. Forbes, and K. Okruhlik, eds., *Proceedings of the 1992 Biennial Meeting of the Philosophy of Science Association*, vol. 2, pp. 13–22. East Lansing, MI: Philosophy of Science Association.

Rudner, R. 1953. "The Scientist *Qua* Scientist Makes Value Judgments." *Philosophy of Science*, 20, pp. 1–6.

Suruda, A. J., E. M. Ward, and M. A. Fingerhut 1993. "Identification of Soft Tissue Sarcoma Deaths in Cohorts Exposed to Dioxin and Chlorinated Naphthalen." *Epidemiology*, 4, pp. 14–19.

PART III

VALUES AND GENERAL
PHILOSOPHY OF SCIENCE
PERSPECTIVES

SEVEN

IS LOGICAL EMPIRICISM COMMITTED
TO THE IDEAL OF
VALUE-FREE SCIENCE?

John T. Roberts

7.1 Introduction

It is obvious that science cannot be completely "value free," just because it is a human activity carried out with certain purposes and goals.
Doing science means adopting those purposes and pursuing those goals,
and doing that, presumably, involves implicitly accepting value judgments. So science cannot be pursued without accepting value judgments.
A *completely* value-free science is an illusion. In my title, though, by
value-free, I don't mean free of *all* values. Rather, value-free here means
free of all value judgments other than epistemic ones, such as the judgment that true theories, informative theories, or empirically adequate
theories would be good things to have. Nonepistemic values are those implicated in value judgments we adopt for nonepistemic reasons. They include, presumably, moral, political, and social values and perhaps more
besides. So my question is whether logical empiricism is committed to
the proposition that a science pursued without regard to any values except epistemic ones is an ideal we ought to pursue.

    This sounds like a historical question, but that is not how I intend
it. My question is not whether those philosophers whom we call "the
logical empiricists" were committed to the ideal of value-free science

but rather whether that approach to the philosophy of science that we call "logical empiricism" is so committed. Whether, say, Carnap or Hempel or Reichenbach actually believed that science ought to be value free is a separate issue; what I am asking is whether the fundamental philosophical commitments they shared, by virtue of which we call them all logical empiricists, force them to have this belief. Or turning the question around: Could you consistently maintain that the logical empiricist approach in philosophy of science is the right one to take, while denying that science is or even ought to be value free?

It seems obvious that the answer to my titular question is yes. That is, once it is made more definite just what is meant by "the logical empiricist approach to philosophy of science," and it is made more definite just what is meant by "the ideal of value-free science," it seems obvious that anyone committed to the former is, at least implicitly, committed to the latter. Because I put the point in terms of implicit commitment, I should pause to say something about how I am using the term *commitment*.

As I use this term, the following is true:

> If you're committed to X, and Y provides (part of) a justification for X,[1] and there's no way to motivate or justify X if you deny Y, then you're (at least) implicitly committed to Y.

The idea is that you are committed to the only possible justification for what else you're committed to (if there is a unique possible justification for it). Of course, there are perfectly good senses of "commitment" in which this isn't true. But this is how I'm going to use the term. So, the view that I just said seems obvious is that the proposition that we ought to pursue the ideal of a (nonepistemic) value-free science can provide part of a justification for the logical empiricist approach, and that there seems to be no way of motivating the logical empiricist approach if you reject that ideal.

Again, I think that seems obvious, but in this chapter I'll argue that it is actually false. In fact, the argument I present will, if it works, show that the actual sate of affairs is the diametrical opposite of the apparent one: Logical empiricism is (implicitly) committed to rejecting the ideal of value-free science. (I'm not completely sure that the argument works, but I think it's at least worth giving a run for its money.)

In presenting this argument, I don't mean to give a reason to endorse or to reject logical empiricism. That approach to philosophy of science has been subject to numerous impressive criticisms, many of which have nothing at all to do with the topic of values in science. If being committed to the ideal of value-free science is a bad thing, then showing that logical empiricism is not committed to that value is hardly enough to vindicate it. What I aim to do here is just to map out a certain region of logical space, in which reside various competing ideas about how logical empiricism could best be motivated. If the argument I'm going to present here works, though, I think it may have some relevance for contemporary philosophy of science, because it yields a general lesson about the possible relations among nonepistemic values, science, and philosophy of science.

The structure of what follows is this: In section 7.2, I will say a little about what I mean by "the logical empiricist approach to philosophy of science." In section 7.3, I will say more precisely what I mean by "the ideal of value-free science." In section 7.4, I will return to the logical empiricist approach to philosophy of science and fill in a few more details. In section 7.5, I will consider and reject an argument that seems to show that what I am calling the logical empiricist approach to philosophy of science must be committed to what I am calling the ideal of value-free science. In section 7.6, I will present the main argument of this paper. Concluding remarks will be offered in section 7.7.

7.2 The Logical Empiricist Approach to Philosophy of Science, Part 1

The logical empiricists (here I'm thinking primarily of Carnap, Reichenbach, and Hempel) saw their work as continuous with the work of scientists. But they didn't think they were doing science itself. According to them, the products of their work were not statements with any cognitive meaning. (Hence, their work was not science.) Rather, what they produced were proposals for how to do various things. For example, they produced proposals for how to understand what the meaning of a scientific theory is, under what circumstances to count a given body of evidence as confirming a given hypothesis to such-and-such a degree, and under what conditions to deem a putative explanation a good one. These are practical proposals concerning how to act—in particular, how to act while doing science. Such proposals, for the logical empiricists, have no cognitive content; they cannot be evaluated as either true or false. Rather, they have to be evaluated in terms of how likely they are to serve some interest of ours.[2]

Because logical empiricist philosophy of science is not in the business of making claims that could be true or false, it could not be a descriptive enterprise. This suggests the familiar idea that for the logical empiricists, philosophy of science is a normative enterprise. This familiar idea is in a sense correct, but it can be misleading if taken to mean that a logical empiricist philosopher of science is a judge for evaluating the work of scientists. Rather, logical empiricism aims to produce new conceptual tools, in the form of new proposals for how to create new science. Alan Richardson (2000) has a helpful way of putting the point:

> Rather than viewing themselves as judges or coaches for scientists, Carnap and Reichenbach were, in the first instance, interested in providing tools for understanding and, thus, doing science. . . . In a Carnapian conceptual engineering project for science, the conceptual tools of current science are used to help create even more precise conceptual tools. (Richardson 2000, S161)

Following Richardson's lead, I will here understand the logical empiricist approach in philosophy of science as a way of going about designing new conceptual tools for the production of future science.

7.3 The Ideal of (Nonepistemic) Value-Free Science

If what you are trying to do is design tools for the production of future science, then it makes sense to ask what science is *for*. That at least a lot of science is meant to be applied is uncontroversial (even for the logical empiricists, whose primary interest usually seems to be in pure science). Applied science is apparently *for* promoting our nonepistemic values: doing things like relieving physical suffering and increasing material flourishing. So if you want to design the best conceptual tools for scientists that you can, then you ought to try to design tools that will help to serve these nonepistemic values. I suppose that few people would deny that in *that* sense, science is not and ought not to be (nonepistemic) value free; certainly, a reasonable logical empiricist would not.

However, there is a widespread and fairly commonsensical view, which I am going to call "the Commonsensical View," according to which, in a different sense, science ought to be value free, for though much science serves, and ought to serve, our nonepistemic values, it can do so effectively only when it does so *indirectly*. The direct goal of science is only to serve our epistemic goals.

> *The Commonsensical View*: Science ought to aim only at serving our epistemic values (like truth, empirical adequacy, informativeness), not our nonepistemic (social, moral, political) values. It is by faithfully serving the former that it can, indirectly, serve the latter. And so, if you're designing conceptual tools for the production of future science, then you ought to keep your eye on the goal of making tools that serve our epistemic values, without getting distracted by the idea that you want to make tools that will serve our nonepistemic values.

A cute analogy is provided by the case of hammers. One of the most important things hammers are for is to help produce sturdy houses. But if you're a hammer engineer, designing the hammers of the future, then you should focus only on serving the value of effective nail driving. It is by serving this value that good hammers can, indirectly, serve the greater value of having sturdy houses. If you keep worrying about how to make a hammer that will best serve the value of sturdy houses, then you're just going to get distracted. In the case of designing the conceptual tools for building the science of the future, there is an even greater need to keep your eye on the direct goal, that of serving our epistemic values; trying to make tools that will serve nonepistemic values not only will be a distraction but also could have a positive pernicious effect (as is illustrated by the examples of the Lysenko affair, Nazi condemnation of "Jewish science," creation science, and all the usual bugbears).

However, there is an important objection to the Commonsensical View. A lot of actions that scientists perform in the course of their scientific activity are practical actions with direct consequences of moral or social significance. Obvious examples include experimentation on human and animal subjects and research applicable to the design of weapons of mass destruction. Another type of example is provided by decisions to accept or reject hypotheses about the effectiveness and safety of a particular drug. In such decisions, getting a false positive and getting a false negative typically have differentially harmful effects. So it makes sense to design decision procedures for deciding whether to accept or reject such hypotheses in a way that gets risk assessment into the picture. But risk assessment requires direct reference to our moral and/or social values: Would it be worse to accept the hypothesis that the drug is safe and effective, even though it's not, and cause harm (or at least no help) to the people who end up taking it (though making lots of money for the drug industry)? Or would it be worse to accept the hypothesis that the drug is not both safe and effective, when in fact it is, and thus delay the marketing of a product that otherwise would have helped many people (and increased profits)? To answer such questions, we need to take account of our nonepistemic value judgments. So, there is a large and interesting range of cases where a conceptual tool for the production of future science ought to be designed in a way that takes account of our nonepistemic values. Cases in which nonepistemic values become directly relevant to scientific practice via risk assessment are discussed in Hempel's "Science and Human Values" (1960). Such examples are discussed in greater scope and detail in Douglas (2000).

This point looks like an important objection to the Commonsensical View: It seems clear that at least in a great many cases, the procedures and conceptual tools of science ought to be designed in a way that takes our nonepistemic values into account directly. But the objection is not fatal to the spirit of the Commonsensical View, for there's a way of modifying the view that seems to get around the objection:

> The Sophisticated Commonsensical View: Science ought to aim only at serving our epistemic values (like truth, empirical adequacy, informativeness), not our nonepistemic (social, moral, political) values, *except insofar as a given scientific action has practical consequences of direct moral or social significance*. It is by faithfully serving the former that it can, indirectly, serve the latter. And so, if you're designing conceptual tools for the production of future science, then you ought to keep your eye on the goal of making tools that serve our epistemic values, without getting distracted by the idea that you want to make tools that will serve our nonepistemic values, *except insofar as the employments of those tools will be actions with practical consequences of direct moral or social significance*.

I take it that what we might call "the received view" of logical empiricism has it that logical empiricism is committed to the Commonsensical View—except when it's in the hands of someone like Hempel, who is very sensitive to the sorts of issues just raised, and then it is committed to the Sophisticated Commonsensical View. Both the Commonsensical View and the Sophisticated Commonsensical View imply that science ought to be autonomous of our nonepistemic values in one sense but should be required to respect them in another sense. It should *respect* those values in the sense that it is, at least in part, *for* serving those values. But it ought to be *autonomous* of them in the sense that its business is, in the first instance, to try to serve our epistemic values, so it ought not to take nonepistemic values directly into account (*except maybe* in those cases where a scientific action has important practical consequences that need to be addressed directly). By analogy, hammers are in one sense not autonomous of the value of sturdy houses because one of the things they're for is helping to serve that value. But in another sense, hammers ought to be autonomous of the value of sturdy houses because a hammer's proper job is just driving nails, and in designing hammers you ought to take into account only the value of effective nail driving, not the value of sturdy houses (*except maybe* in those cases where the employment of a hammer would have direct practical consequences that are incompatible with our promoting the production of more sturdy houses—for example, when a proposed hammer design would drive nails by means of the force from a thermonuclear blast).

What I mean by "the ideal of value-free science" is just what the Commonsensical View and the Sophisticated Commonsensical View have in common: the idea that science ought to be autonomous of nonepistemic values in the sense of autonomy just described. So the design of conceptual tools for the production of future science ought not to take nonepistemic values into account (except maybe insofar as practical consequences of direct moral or social significance need to be considered).

7.4 The Logical Empiricist Approach to Philosophy of Science, Part 2

What is most distinctive about logical empiricism as an approach to philosophy of science is, first, its denial of synthetic a priori scientific knowledge and, second, its attempt to use the tools of modern formal logic to articulate its proposals, following the lead of early-twentieth-century logicist philosophy of mathematics. For a logical empiricist, then, scientific hypotheses and theories have to be justified by appeal to empirical evidence, and because evidence must be understood in a way that lets formal logic get a grip on it, it must be understood in terms of formaliz-able evidence statements, or observation statements. The proposals sought by logical empiricists were, then (at least for the most part), articulated solely in terms of formalizable observation statements, theories or hy-potheses themselves understood as formalizable statements, and formal, logical relations among all these statements. Thus, Carnap proposed to understand the meaning and meaningfulness of a scientific hypothesis in terms of its formal relations with (true or false) observation statements; Hempel and Oppenheim sought a proposal for evaluating putative ex-planations in formal terms; the various logical empiricists proposed ways of evaluating the degree of evidential support given a hypothesis by a body of evidence in terms of logical relations among hypotheses and ob-servation statements. Perhaps not all of the work in philosophy of sci-ence done by logical empiricists can be neatly fit into the pigeonhole I've proposed. That is, perhaps some of it does not take the form of pro-posals articulated in terms of statements and formal relations among statements. But if one wants to abstract away from the details of the work of particular logical empiricists and talk about "the logical empiri-cist approach to philosophy of science," then this seems as fair a way of doing it as any.[3]

In what follows, I am going to focus on one strand within the logi-cal empiricist tradition: the effort to articulate a good proposal for evalu-ating the degree of evidential support afforded a hypothesis by a given body of evidence. I'll call any such proposal a "confirmation theory." (Of course, there are other senses in which one could use that term.) Not all confirmation theories are logical empiricist ones, of course. What makes a confirmation theory a logical empiricist one is that it under-stands a given body of evidence as a given set of observation statements and that the proposal refers to no features of the hypothesis or observa-tion statements other than their formal, logical relations to one another. By this standard, Carnap's (1950) theory of logical probability counts as a logical empiricist confirmation theory. Hempel's (1945a, 1945b) quali-tative theory of confirmation seems to fit the bill as well. Hempel does allow such nonformal things as simplicity to play a role in theory choice, but it seems a safe bet that if Hempel could have formalized simplicity, then he would have been happy to do so. So it seems fair to say that Hempel's goal was a proposal that would fit my proposed defi-nition of a logical empiricist confirmation theory.

There are plenty of nonempiricist confirmation theories one could adopt. For instance, it could be argued with some plausibility that the seventeenth-century mechanical philosophers (implicitly) adopted a confirmation theory according to which certain features of the *content* of a hypothesis, over and above its logical form and its formal relations to observation statements, are relevant to assessing its degree of support. On this confirmation theory, how well a hypothesis is supported by evidence depends, among other things, on whether the hypothesis posits any kind of causality other than contact action. If it does, then it can be ruled out immediately and isn't supported by any evidence, no matter what the logical relations between it and this evidence may be. More generally, if you think we have some synthetic a priori knowledge of the natural world and that any empirical theory must be consistent with this a priori knowledge to be empirically defensible at all, then you aren't a logical empiricist, for you hold that the relation of evidential support depends on something other than logical relations among hypothesis and evidence statements. Another kind of alternative to the logical empiricist approach to confirmation theory is the proposal that evidential support in science be assessed according to the rule of inference to the best explanation, where it is stipulated that the goodness of an explanation is not a formal matter. On such a proposal, what makes a hypothesis well supported by a set of observation statements is the obtaining of certain relations between these statements that are not wholly formal or logical but depend on some feature of the contents of the hypothesis and the observation statements.

The availability of these alternatives, and the ways in which they differ from the confirmation theories proposed by logical empiricists, illustrate that what I'm calling the logical empiricist approach to confirmation theory is an important distinctive feature of the logical empiricist tradition. In what follows, I'm going to focus on confirmation theory to the exclusion of other aspects of the logical empiricist program. Our question, then, becomes more focused: If you are committed to a logical empiricist confirmation theory (either to *some particular* such theory or to the goal of finding an acceptable such theory), then are you (at least implicitly) committed to the ideal of value-free science? I focus here on confirmation theory because it will make the argument to follow smoother. But I think a similar argument could be given focusing on any of the other types of proposal, mentioned previously, sought by the logical empiricists, such as proposals for how to decide when a hypothesis is meaningful. The more focused question seems to have an obvious answer (namely, "Yes!"). In the following section, I'll reject an apparently plausible argument for this answer, and in the next section, I'll present a positive argument against this obvious answer.

7.5 A Quick Argument to Show That Logical Empiricism Is Committed to Value-Free Science (at Least in the Case of Confirmation Theory)

I've suggested that being a logical empiricist means being committed to articulating and advocating proposals for evaluating the degree of evidential support of hypotheses by using criteria that are specified solely in terms of formal relations among hypotheses and observation statements. If this is so, then it seems that being a logical empiricist must mean being committed to the idea that science ought to be designed in a way that does not take nonepistemic values into account, for our nonepistemic values are surely not going to show up in the formal properties of hypotheses and observation statements. But that means they'll be ignored by a logical empiricist confirmation theory! More carefully:

1. In her confirmation theory, a logical empiricist proposes criteria for deciding how worthy of acceptance a given hypothesis is in the light of a given set of observation statements.
2. Those criteria make no mention of anything but formal, logical stuff—no room for any considerations of nonepistemic values.
3. So, a logical empiricist is committed to the proposal that the evaluation of the evidential support of hypotheses should be insensitive to nonepistemic values. (From 1 and 2)
4. So, a logical empiricist is committed to the proposal that (a central part of) scientific activity be (nonepistemic) value free. (From 3)

This argument seems convincing at first glance. But we should be wary of it, because it proves too much. True, the criteria that are granted admission in a logical empiricist confirmation theory do not include any nonepistemic values. But they don't include any of the paradigmatic epistemic values—truth, informativeness, empirical adequacy[4]—either! If this argument proves that a logical empiricist must be committed to the autonomy of science from our nonepistemic values, then a parallel argument proves that a logical empiricist is committed to the autonomy of science from the values of truth, empirical adequacy, and informativeness as well.

Clearly something has gone wrong. I offer the following diagnosis: The inference from 1 and 2 to 3 is not valid. From the lack of any mention, in criterion C, of value V, it cannot be inferred that criterion C was not designed to serve value V, or that someone proposing criterion C could not consistently do so because of a desire to serve or respect value V. For example, the legal rules of evidence lay down criteria for deciding when a piece of evidence may and may not be admitted into a trial. These rules (presumably) don't ever mention the value of treating the accused in a just manner. But it does not follow that those who design or endorse these rules may not do so precisely because they want to respect the value of treating the accused in a just manner. Nor does it follow that they are committed to the idea that the criminal court system ought to be autonomous of the value of justice. Similarly, just because a philosopher seeks a confirmation theory that articulates criteria for assessing degree of support that nowhere mention nonepistemic values, it does not follow that this philosopher is committed to the idea that science ought to be autonomous of our nonepistemic values. It seems to me that this observation undermines the most plausible apparent reason to think that a logical empiricist must be committed to the ideal of value-free science.

7.6 The Main Argument

Suppose that the question before us is whether we ought to pursue confirmation theory in the logical empiricist way at all. This is distinct from the question whether we ought to adopt some particular logical empiricist confirmation theory (though I'll return to that question later). Commitment to logical empiricism, of course, is not a commitment to any particular set of tenets but rather to a certain way of doing philosophy of science. I suggest that a useful way of understanding this commitment is as a commitment to a *metaproposal*: a proposal for how to formulate and evaluate proposals for how to do various things (how to understand the meaning and meaningfulness of scientific hypotheses, how to evaluate degree of evidential support, etc.). In the area of confirmation theory, the logical empiricist metaproposal can be formulated as follows:

The Logical Empiricist Metaproposal for Confirmation Theory: Seek
to develop criteria, for deciding to what extent a given hypothesis is
supported by a given set of observation statements, that are articulated
solely in terms of formal, logical relations among hypotheses and observation statements.

Our question now is how to justify this metaproposal. As a proposal, it
has no cognitive content (on *any* logical empiricist criterion of cognitive content). To justify this metaproposal, then, is not to find reasons
to think it is true but rather to find reasons to think that adopting it is
likely to be beneficial. In this section, I will argue, first, that a logical
empiricist who is committed to the ideal of value-free science (as I
have formulated this ideal in section 7.3) is unable to give any complete and coherent justification of this metaproposal and, second, that
a logical empiricist who is not committed to that ideal may well be in a
position to offer such a justification. For this reason, I will conclude
that logical empiricism as such is not committed to the ideal of value-
free science.

Suppose a logical empiricist accepts what I am calling the ideal of
value-free science, that is, what is common to the Commonsensical View
and the Sophisticated Commonsensical View. On this basis, it seems that
any justification of the logical empiricist metaproposal for confirmation
theory must involve the following claim, which I'll call VF:

VF: We ought to adopt the logical empiricist metaproposal, because it is
more likely than its alternatives to produce conceptual tools that
will serve our epistemic values (e.g., truth, empirical adequacy, informativeness) effectively.

VF has a component that, for a logical empiricist, is without cognitive
content,

We ought to seek true (or at least empirically adequate), informative
theories

and a component with cognitive content,

Adopting the logical empiricist metaproposal is more likely than the alternatives to produce tools that serve the goal of finding true (or at least empirically adequate) and informative theories.

If VF is to be used in a plausible justification of the logical empiricist metaproposal, then it seems that the component of VF that has cognitive content must be given some justification as well. This component is a very strong synthetic claim about the world. So for a logical empiricist, it requires empirical justification. But how can a logical empiricist offer such a justification? What is needed is empirical evidence that adopting a confirmation theory articulated solely in formal terms is more likely to lead to true (or empirically adequate), informative theories than is adopting any alternative confirmation theory, such as one that depends on a nonformal notion of explanation or one that demands that a hypothesis not be considered supportable at all unless it is consistent with certain alleged items of synthetic a priori knowledge. Even if a logical empiricist is in a position to reject arguments for the existence of synthetic a priori knowledge, that is not sufficient here. For what is needed is an empirical justification for the claim that adopting a confirmation theory that does *not* demand consistency of hypotheses with, for example, Euclidean geometry or the nonexistence of action at a distance is more likely to help us get true or empirically adequate hypotheses than one that does. If one of the alleged synthetic a priori principles turns out to be true, or at least empirically adequate, then a confirmation theory demanding consistency with this principle might well be more likely to serve our epistemic values than one that doesn't, even if we lack any faculty that allows us to know these principles a priori. So the burden one must carry if one is to justify VF is not the burden of rejecting apriorist epistemology; it is, rather, the burden of finding evidence for the very strong empirical component of VF.

What could a logical empiricist offer as a good reason to believe that the empirical component of VF is true? It seems that such a reason could be given only on the basis of a great deal of substantive, empirical knowledge about the nature of the world. We would be prepared to offer such a reason only if we already had a much better-developed science than we now have. So it seems that at the time when such a reason is needed—namely, now, while the task of producing the science of the future still confronts us—it isn't available. By their own lights, then, logical empiricists have no evidence to point to that can justify the empirical component of VF. In most general terms, the problem facing the logical empiricist here is that of finding empirical evidence to justify an extremely strong synthetic claim, which must be justified before we can justify the logical empiricist metaproposal concerning confirmation theory. It appears that such evidence is unavailable before the task of science is complete, or at least is very much further along than it is at present. I'll call this problem *the synthetic obstacle problem*, because it involves an obstacle to the justification of the logical empiricist approach that takes the form of a synthetic claim needing justification.

It might seem that the synthetic obstacle problem is nothing more or less than the problem of induction. If that were so, then one might be tempted to say that the inability of a logical empiricist to solve this problem is neither surprising nor troubling. If the task of justifying the logical empiricist metaproposal—and hence the logical empiricist approach to confirmation theory—on the basis of the ideal of value-free science has been reduced to the problem of induction, then perhaps we ought to say that logical empiricists committed to value-free science are just as able to defend their view as is anyone with a positive view about confirmation theory. However, I think we should reject this tempting thought, for two reasons. The first is that, as I will argue later, there is at least one position on confirmation theory that is more defensible than that of the logical empiricist committed to the ideal of value-free science (for this position does not face the synthetic obstacle problem). Hence, the logical empiricist committed to value-free science is *not* in just as good a position as anyone else. The second is that the synthetic obstacle problem is not identical to the problem of induction, for being able to solve the former is not sufficient for being able to solve the latter. A solution to the synthetic obstacle problem would give us a reason to believe that some logical empiricist confirmation theory is more apt to serve our epistemic values than is any nonlogical empiricist confirmation theory. But it doesn't follow from this that the application of any particular confirmation theory is very likely at all to lead us to the truth. Solving the problem of induction, however, requires showing that some inductive method is likely to lead to the truth.

I've introduced the synthetic obstacle problem as a problem faced by anyone committed to the ideal of value-free science who wants to justify the logical empiricist metaproposal. But it is clear that the same problem will also afflict anyone committed to the ideal of value-free science who seeks to justify some *particular* logical empiricist confirmation theory. The claim that we ought to adopt some particular logical empiricist confirmation theory (say, one of Carnap's inductive systems) also has a component with empirical content, namely, that the adoption of this confirmation theory is more likely to promote our epistemic values than is any of the available alternatives, including all the ones that are inconsistent with the logical empiricist metaproposal. Because this empirical claim is strictly stronger than the empirical component of VF, the synthetic obstacle problem is faced again.

Is there any available way of justifying the logical empiricist metaproposal that does not bring us up against the synthetic obstacle problem? There is, *if* one is willing to give up the ideal of value-free science (and thus reject the Commonsensical and Sophisticated Commonsensical Views). If one gives up that ideal, one is free to seek a "value-committed" justification of the logical empiricist metaproposal, based on the following claim:

VC: We ought to adopt the logical empiricist metaproposal, because doing so directs us, in a way that the available alternatives do not, to seek to develop and advocate methods whose use respects (some of) our nonepistemic values.

How might VC be justified? One way is as follows: Consider what a logical empiricist confirmation theory *excludes* from consideration in deciding how to assess evidential support. Everything except the set of formal, logical relations holding among hypotheses and observation statements gets ruled out. That implies that who thought of the hypothesis and what social groups he or she may belong to are of no matter. This embodies a kind of egalitarianism. But of course, logical empiricist confirmation theories are hardly unique in this regard. It is hard to think of a confirmation theory that has been seriously proposed in recent times that does not share this feature.

However, what is unique about logical empiricist confirmation theories is that they exclude all considerations about the *content* of a hypothesis (as opposed to its logical form and its formal relations to observation statements). This embodies a different kind of egalitarianism: an egalitarianism at the level of views about the way the world is. In turn, this embodies a kind of egalitarianism at the level of cultural traditions and Weltanschauungen (just because different cultural traditions and Weltanschauungen favor different hypotheses over others, on the basis of their contents). More generally, the exclusive focus on formal, logical relations denies privilege to any hypothesis on the basis of its consistency or affinity with any particular conception (or prejudice about) the nature of the natural world. So an egalitarianism at the level of substantive ideas about the world is embodied in a logical empiricist confirmation theory.

Logical empiricists, then, might defend their approach to confirmation theory by pointing out that, unique among types of confirmation theory, logical empiricist confirmation theories embody a certain value—egalitarianism among views of the world. This claim does not involve an empirical component. It involves the claim that we ought to value a certain kind of egalitarianism and the claim that a logical empiricist confirmation theory, as such, exemplifies this kind of egalitarianism. Neither of these is an empirical claim that must wait on the progress of science for its justification. So the synthetic obstacle problem does not arise.

Is the kind of egalitarianism involved here an epistemic or nonepistemic value? Well, just because it is a kind of egalitarianism, it has the ring of a social or cultural value. At any rate, it is not one of the standard paradigms of epistemic values (such as truth, empirical adequacy, and informativeness). One might insist that it really is an epistemic value, because it is nothing more or less than a lack of bias or prejudice about how the world is. However, such a lack of bias seems to be an epistemic value precisely insofar as it seems likely to remove hindrances in finding out the truth about the world. Thus, it seems to be "indirectly" epistemically valuable: valuable as a means to the end of getting the kind of knowledge we want or as a means of clearing away obstacles to our getting the kind of knowledge that we want. But in the value-committed justification of the logical empiricist metaproposal just presented, egalitarianism at the level of views of the world is not held up as something valuable because it is likely to help us achieve our goal of finding out how the world is. Rather, it is held up as something valuable in itself. If it weren't for this, then the proposed value-committed justification of logical empiricism would obviously face the synthetic obstacle problem. (For all we know now, certain systematic biases might actually be helpful in getting us to the truth about the world! Ruling this out would require some impressive empirical knowledge.) So this justification of logical empiricism is advantageous only because it appeals to a certain kind of egalitarianism as something valuable *in itself*. Hence, it is not so clear that this egalitarianism is playing the role of an epistemic value, at least as epistemic values are traditionally conceived.

Note that the proposed value-committed justification of logical empiricism *does not* suggest that adopting a logical empiricist proposal for confirmation theory will tend to bring about greater egalitarianism and toleration in our society, or anything like that. If that were the suggestion, then it would be an empirical, social-scientific claim, and we would not be in a position to offer it until we already had on hand a significant amount of social-scientific knowledge. So once again we would be face to face with the synthetic obstacle problem. Rather, the suggestion is that the adoption of a logical empiricist confirmation theory would *embody* a certain thing we value (for apparently nonepistemic reasons), because doing science in accord with the proposal made by such a confirmation theory *would itself be an instance of the thing valued*; it would itself be an instance of a certain sort of egalitarian decision making. This is why the synthetic obstacle problem is skirted. Unlike VF, VC does not have a component with cognitive content that would require a tremendous amount of empirical knowledge about the world in order to justify it. All that is required is that we do, in fact, value the kind of egalitarianism at issue, which any logical empiricist confirmation theory manifestly instantiates.

Of course, one might just say that the kind of egalitarianism at issue is not something that we should value. Here I can offer no argument that we should value this kind of egalitarianism. At this stage of the debate, a logical empiricist is likely to say that we have reached the place where rational justification has run out; all we can say is that we do, or do not, find egalitarianism at the level of general views of the world valuable, and in doing this we are not making claims with cognitive content that could be justified by appeal to evidence. Again, my aim here is not to vindicate logical empiricism but to explore the prospects for offering a coherent justification of it. Even if one denies the value of the kind of egalitarianism at issue here, it still seems that the proposed justification is coherent and that it goes as far as one can go by logical empiricist lights. This is what we can't get if we try to justify logical empiricism on the basis of the ideal of value-free science.

I've looked at two different ways in which one might try to motivate the logical empiricist project in the first place—or at least a big part of that project (confirmation theory). One of these motivations involves commitment to the ideal of value-free science: the idea that science ought to operate in a way that is, in an important sense, autonomous of our nonepistemic values, and so new proposed methods and conceptual tools for science can be justified only in a way that makes no appeal to those values. The other motivation involves commitment to the idea that science ought not be autonomous of values: that scientific activity ought to embody a certain social value and that the procedures and conceptual tools of science ought to be designed with that value in mind. If the argument I've given works, then the second motivation has at least a chance of succeeding, whereas the first motivation has none.

But between these two proposals for motivating the logical empiricist project, the difference that makes a difference is not really that between epistemic and nonepistemic values. Rather, it's a difference in the way that values function in the justification of a certain way of doing philosophy of science. The (nonepistemic) value-free justification of the logical empiricist metaproposal says that we should adopt that metaproposal because doing so will lead us to develop tools that will help us get something we want (namely, informative true theories, or at least informative empirically adequate theories). The (nonepistemic) value-committed justification says that we should adopt that metaproposal because doing so will lead us to develop tools whose very use will instantiate something that we value, not that doing so will enable us to get something else that we value. The crucial difference here is not between epistemic or nonepistemic values but rather between *consequentialist* and *deontological* justifications. It is because the value-free justification seeks to justify the logical empiricist metaproposal in consequentialist terms—"we should adopt that metaproposal because it is the best available way to get what we want"—that it requires that we be in a position to offer evidence for a very strong synthetic claim, the claim that one alternative among many others is the most effective means to our end. The alternative, deontological proposal does not face this objection. Someone who tries to justify the logical empiricist metaproposal on the grounds that this proposal directs us to develop tools whose very use will be an instance of something we value does not commit herself to any synthetic claims about what the most effective means to a given end is. She does stick her neck out in a different sense, by claiming that a certain kind of egalitarianism is something we ought to value. But perhaps any justification of any practical proposal, in science or philosophy of science, must ultimately rest on such a claim.

7.7 Conclusions

If the argument of section 7.6 succeeds, then it establishes two important conclusions:

1. There is a coherent way of motivating the logical empiricist approach to philosophy of science that works by pointing out that this approach directs us to design conceptual tools that *directly* embody one of our nonepistemic values (*rather than* indirectly serving our nonepistemic values as a means to the end of finding true, or empirically adequate, informative theories).
2. The more obvious way of motivating the logical empiricist approach, by arguing that adopting that approach will help us achieve our cognitive goals of getting true, empirically adequate, informative theories, faces a very serious difficulty (in the form of the synthetic obstacle problem) that is not faced by the alternative mentioned previously.

Note that the second conclusion does not imply that our epistemic values have no role to play in logical empiricist philosophy of science. Our epistemic values might well be relevant to the question whether one logical empiricist confirmation theory ought to be preferred over another.[5] They might also be relevant to various other questions, such as which formal relations among statements are the important ones or which statements should be considered observation statements.[6] The second conclusion says only that an appeal to our interest in promoting our epistemic values (truth, informativeness, empirical adequacy) cannot alone justify adopting the logical empiricist approach.

Given the stipulation about implicit commitment made in section 7.1, the first of these conclusions has the upshot that logical empiricism is not committed to the ideal of value-free science. The second conclusion implies, further, that logical empiricism is committed to rejecting that ideal. So if the argument of section 7.6 fully succeeds, then the actual state of affairs with respect to the logical empiricist approach to philosophy of science and the issue of value-free science is strikingly different from the apparently obvious state of affairs.

A more general upshot of this argument is that although any way of designing tools for science can be justified only on the basis of some value judgment or other, the justification doesn't have to take the form:

> X is desirable, so science should be designed and done in such a way as to promote the attainment of X.

In fact, there are serious difficulties facing any such justification. A different form of justification, which doesn't face the same difficulties, is:

> X is desirable, so science should be designed in such a way that its practice will instantiate or embody X.

A justification of the second form could appeal to an X that is valuable for either epistemic or nonepistemic reasons. But the conventional way of thinking of science as motivated only by epistemic values seems difficult to put into the second form. According to that conventional way of thinking, we ought to do science in one way rather than another because that way is the best one for serving our interest in getting things that we want, namely, theories that have certain desirable features. The focus is on qualities desired in the *product* of scientific inquiry. A justification of the second form, by contrast, requires focusing on qualities desired in the *process* of inquiry—and *not* qualities in the process that are desired solely as a means to producing a valuable product.

ACKNOWLEDGMENT

I would like to thank Gerald Doppelt, Uljana Feest, Livia Guinaraes, Thomas Hill Jr., Janet Kourany, Roger Sansom, and especially Geoffrey Sayre-McCord for valuable comments and discussion.

NOTES

1. By "justification," here I mean hypothetical justification: If Y itself is justified, then it can provide (part of) a justification for X. This doesn't require that Y itself be justified for Y to provide (part of) a justification for X.

2. An example of this view is provided by Carnap in "Testability and Meaning," in which Carnap defends a logical empiricist view of how the meanings of hypotheses should be understood ("the principle of empiricism"):

It seems to me that it is preferable to formulate the principle of empiricism not in the form of an assertion—"all knowledge is empirical" or "all synthetic sentences that we can know are based on (or connected with) experiences" or the like—but rather in the form of a proposal or requirement. As empiricists, we require the language of science to be restricted in a certain way; we require that descriptive predicates and hence synthetic sentences are not to be admitted unless they have some connection with possible observations, a connection which has to be characterized in a suitable way. (Carnap 1937, 33)

3. Again, my primary aim here is not a historical one.

4. Here and elsewhere in this chapter, by "empirical adequacy" I mean not just adequacy in the light of empirical evidence on hand at present but adequacy to all the empirical phenomena.

5. One obvious example: Our interest in finding true theories can motivate one to prefer, say, Hempel's qualitative theory of confirmation to one that assigns high confirmation only to hypotheses that are formally inconsistent with at least one observation statement.

6. Thanks to Gerald Doppelt for helpful discussion that made me realize the importance of this.

REFERENCES

Carnap, R. 1937. "Testability and Meaning—Continued." *Philosophy of Science*, 4, pp. 1–40.

Carnap, R. 1950. *Logical foundations of probability.* Chicago: University of Chicago Press.

Douglas, H. 2000. "Inductive Risk and Values in Science." *Philosophy of Science*, 67, pp. 559–79.

Hempel, C. G. 1945a. "Studies in the Logic of Confirmation (I)." *Mind*, 54, pp. 1–26.

Hempel, C. G. 1945b. "Studies in the Logic of Confirmation (II)." *Mind*, 54, pp. 97–121.

Hempel, C. G. 1960. "Science and Human Values," in R. E. Spiller, ed., *Social Control in a Free Society*, pp. 39–64. Philadelphia: University of Pennsylvania Press.

Hempel, C. G., and P. Oppenheim. 1948. "Studies in the Logic of Explanation." *Philosophy of Science*, 15, pp. 135–75.

Richardson, A. 2000. "Science as Will and Representation: Carnap, Reichenbach, and the Sociology of Science." *Philosophy of Science*, 67, pp. S151–62.

EIGHT

CONSTRUCTIVE EMPIRICISM
AND THE ROLE OF SOCIAL
VALUES IN SCIENCE

Sherrilyn Roush

> To accept a theory is to make a commit-
> ment, a commitment to the further con-
> frontation of new phenomena within
> the framework of that theory . . . and a
> wager that all relevant phenomena can
> be accounted for without giving up that
> theory. Commitments are not true or
> false; they are vindicated or not vindi-
> cated in the course of human history.
> —Bas C. van Fraassen,
> *The Scientific Image*

IN HER BOOK *SCIENCE AS SOCIAL KNOWLEDGE,* HELEN LONGINO ARGUED
not only that social values are in fact ineliminable from theory choice in
science but also that we ought to rewrite our ideals in such a way as to
incorporate this fact. One of the most common criticisms one hears of
this idea of granting a legitimate role for social values in theory choice
in science is that it just doesn't make sense to regard social preferences
as relevant to the truth or to the way things are. "What is at issue," wrote
Susan Haack, is *"whether it is possible to derive an 'is' from an 'ought.'"*
One can see that this is not possible, she concludes, "as soon as one expresses it plainly: that propositions about what states of affairs are *desirable* or *deplorable* could be evidence that things *are,* or *are not,* so"
(Haack 1993a, 35, emphasis in original). Haack does not provide an argument for the view that it is impossible to derive an *is* from an *ought,*
but the intuition she expresses is strong and widespread. The purpose of

this chapter is not to determine whether this view is correct but rather to show that even if we grant it (which I do), we may still consistently believe that social values have a legitimate role in theory choice in science.

I will defend this conclusion by outlining a view about social values and theory choice that is based on the constructive empiricism (CE) of Bas van Fraassen. Some questions about what role social values may legitimately play in science look different, I contend, depending on whether they are viewed from a realist perspective, according to which the aim of science is literally true description of reality, or from the point of view of a constructive empiricist antirealism, according to which the aim of science is empirical adequacy, the fit of a theory to the observables.[1] In Longino's account of the role of social values in science, she expressed what appears to be a realist view when articulating one of the goals she took science to have:

> My concern in this study is with a scientific practice perceived as having true or representative accounts of its subject matter as a primary goal or good. When we are troubled about the role of contextual values or value-laden assumptions in science, it is because we are thinking of scientific inquiry as an activity whose intended outcome is an accurate understanding of whatever structures and processes are being investigated. (Longino 1990, 36)

Consequently, when Longino went on to argue that social values had to play a role in deciding between theories because one had to choose with insufficient evidence which auxiliary assumptions to adopt in order to have any view about which evidence was relevant to a theory, she appeared to step just where Haack insists we must not. She appeared to commit herself to the view that social preferences could play the role of reasons to believe an assumption was true, or at least could make adoption of an assumption legitimate.

As I develop CE in what I take to be the most natural way toward a view of the role of social values in theory choice, it will become clear that there is a sense in which the view I describe is different from, and not consistent with, Longino's 1990 view. Nevertheless, the conclusion I draw, that social values indeed may have a legitimate role in theory choice, is obviously a defense of part of Longino's overall claim. After defending CE's solution to the problem of social values in theory choice against some obvious objections, I will argue that an attractive realist way of attempting to achieve something similar is not successful. This suggests that CE may be the only way to grant a legitimate role to social values in theory choice without falling prey to Haack's objection.

8.1 Limitations of the Objection

It goes without saying that social values are often involved in the decisions we make, consciously and unconsciously, about which subjects to inquire into and which questions to ask, and often legitimately so. Haack's objection that an *ought* does not imply an *is* does not undermine the legitimacy of social values playing a role in determining *which* things we learn about the world, nor does Haack think it does. But the relevance of Haack's criticism to the sorts of matters those writing about social values and science are concerned about is narrower still. The force of Haack's objection can be maintained only if it is stated pretty much exactly as she has put it: Preferences that a thing be so cannot be evidence that it is so, or a reason to think it is so. There are claims that sound similar but are found on inspection to be indefensible. For example, we might have thought Haack said that the way we want things to be cannot be relevant to the way things are (as I stated the matter in my opening sentence). But this is manifestly false if the "things" referred to are themselves social things, to take the easiest case; our preferences are the reasons many social things *are* as they are, though not alone reasons to *think* they are as they are. And the way we want things to be is clearly to some extent relevant to the way they *will be* in the future, even with nonsocial nature.

We might have thought Haack's objection was the same as saying that social values cannot be *epistemically* relevant. But if it were, then it would be wrong, because everyone knows that social values can introduce bias and that bias is relevant to whether we have good epistemic reasons. Finally, it must surely be acknowledged that the social structure of a community is relevant to how successful its members will be in finding correct answers to their questions. To take an extreme case, if the social structure of a community was such that everyone believed on authority the views of a certain individual rather than investigating any claims on their own, then it would be less likely that this community would find good answers to questions about the world than it is for a community of independent and interacting investigators. Social structures can embody social values that are also epistemic values, though the way they function as epistemic values is in governing practices, not directly as reasons to believe a certain theory is true. Because they have an epistemic role to play, and epistemic questions are constitutive of science on anyone's view, that these sorts of social values have a function in science does not need to be defended by the sort of view I am developing. To summarize, some of the most discussed ways in which social values might be relevant to science do not fall prey to Haack's objection; I will focus on those that do.

8.2 Social Values in Theory Choice

According to the constructive empiricist, accepting a theory need not and should not involve believing that it is true. That is, it need not and should not involve believing that the theory's claims about all types of things—observable and unobservable—are correct. Accepting a theory should involve believing only that it is empirically adequate, that is, believing that it fits all observable phenomena, those we have actually observed and those we have not. The source of the flexibility about theory choice that I find in CE lies in the following two aspects of the view: (1) For a given domain, there is only one true theory, whereas there are in general many empirically adequate ones (whether we can imagine them or not). This follows roughly from the meanings of the terms. (2) On this view, virtues of a theory that go beyond consistency, empirical adequacy, and empirical strength do not concern the relation between the theory and the world; "they provide reasons to prefer the theory independently of questions of truth" (van Fraassen 1980, 88). On this admittedly controversial view, there isn't any kind of evidence we could get that would make it more rational to think a theory was true than to think it was empirically adequate. So, since when we might be wrong we should be wary of making ourselves wrong about more things by believing stronger claims, we are better off restricting our beliefs about theories to claims of empirical adequacy.

On standard assumptions about truth, there is at most one theory that is true of a given domain up to notational variation, and therefore at most one theory that it would be correct to believe, because believing, for van Fraassen at least, means believing true. However, there are many theories about a domain, including many we have not thought of, that we would be within our epistemic rights to accept, in van Fraassen's sense of acceptance, because there are potentially many theories that are empirically adequate, true to all observables, in a domain.[2] For the constructive empiricist, social values can legitimately play a role in grounding choices among theories when these are choices among theories all of which are legitimately believed at a given time to be empirically adequate, because the choice of one among these theories is a pragmatic affair.

Constructive empiricism is not as it stands a methodological view that would tell us how to decide which theories are empirically adequate or how to choose between candidate theories. Nevertheless, it has some framing implications for these questions, because it involves a view about what we ought to be doing when we choose a theory, namely, believing that it is empirically adequate on epistemic grounds and preferring it to any rival in the same domain that we also believe to be empirically adequate, on pragmatic grounds that have nothing to do with correspondence between the theory and the world. It follows, crucially, that if we were faced with rival theories that according to our present evidence, both appeared to be empirically adequate, choosing between them would not be a choice of which theory we should believe to be true. Because we would not be choosing which one to believe true, a fortiori we would not be choosing which one to believe true *on the basis of social values*. Thus, if social values were among the pragmatic grounds we appealed to in choosing among theories we believed to be empirically adequate, we would not be committing the fallacy Haack inveighs against. And on this way of placing social values in the activity of theory choice, the possibility that social values (one sort of pragmatic factor) would be a reason to think a theory empirically adequate does not arise. It is assumed that the criteria for judging empirical adequacy are thoroughly epistemic, as they should be—taking the lesson of Haack's objection—because a theory is empirically adequate just in case what it says about observables is *true*.

It is illuminating to compare this view with Longino's (1990) picture of why social values are ineliminable from theory choice. There, social values enter into our judgment not after all current evidence is tallied in the columns of the appropriate hypotheses but before this tallying and as a precondition.[3] Following the bootstrap model of confirmation, Longino concludes that confirmation is relative to auxiliary assumptions or, in words that even a nonbootstrapper could accept, that it is only by means of auxiliary assumptions that the relevance of evidence to a hypothesis can be judged. To require these auxiliaries to be "directly" confirmed—that is, confirmed without reliance on further auxiliaries—would be, according to Longino, to place unreasonable constraints on science that would disqualify much of the science we admire.[4] It would be feasible, she submits, only for "theories" that expressed nothing more than relationships between observables, and our sophisticated science does not look like that. We should require, she thinks, only that auxiliaries be "indirectly" confirmed, an option that allows the influence of interests and values to enter into theory choice through the choice of auxiliary assumptions (Longino 1990, 51–52).

Constructive empiricism looks like a solution to at least part of Longino's worry. It does not, and according to van Fraassen does not need to, place any restriction on scientific theorizing. Our constructed hypotheses may be as elaborate as we please; elaborate theories can have pragmatic significance, and searching for them may even be our most effective way of producing empirically adequate theories. However, because what we take from theories (and auxiliary assumptions) epistemically is only their observable consequences, their observable consequences are also the only parts of them that need confirmation. And according to Longino, as well as to CE, the observable consequences are more likely to be susceptible to "direct" confirmation than other parts of a theory are. There may be further questions about whether CE can sustain this focus on observables without undermining the rationale for elaborate theories, but maintaining both is the advertised view, and it should seem attractive to anyone who has been impressed by Longino's reflections.

It seems to me that if Longino's claim that we should require only indirect confirmation of auxiliaries can be defended in case the aim of science is truth, then it can also be defended in case that aim is empirical adequacy. Those who aim for empirically adequate theories face the incompleteness of their evidence no less than do seekers after true theories, because our actual observations do not add up to all of the observables. And if the seeker of true theories must appeal to auxiliary assumptions to show that evidence is relevant to a theory, then so must the seeker after theories that are empirically adequate, because the latter is equally obligated to link the theories to observ*ables*. However, because empirical adequacy is *truth* for observables, we cannot accept a claim that social values can be reasons for believing in the empirical adequacy of a theory—on pain of the fallacy I began with—and likewise we cannot accept that they can be reasons for believing auxiliary assumptions (are empirically adequate or wholly true) either. Longino's argument applies to empirical adequacy as it does to truth, but its conclusion is unacceptable if that conclusion is stated as I have just done. In this sense, the account I am sketching is at odds with Longino's (1990) view.

This tension between the view I am describing and what appears to be Longino's view arises from the impression her 1990 discussion creates that social values are being substituted for evidence and the consequent impression that we are relaxing an epistemic standard when we allow social values to play a role in theory choice. However, something of a rapprochement between the two views is possible, partly because, though I think the reading I have given Longino's discussion is the natural one, she does not explicitly commit herself to the view that social values are to be *reasons to believe* auxiliary assumptions. Moreover, there is a constructive empiricist way of fleshing out some of what she does say that escapes that formulation and provides an ersatz version of what she has claimed. Longino has rejected the constructive empiricist option I am presenting, for reasons I will elaborate and make some replies to.[5]

Notice that an auxiliary assumption is a claim about the world, just as a theory is, so constructive empiricists aim for any such assumption they accept to be empirically adequate. If two such assumptions have equal amounts of (incomplete) evidence for the claim that they are empirically adequate, and neither is falsified, then they can legitimately base a choice in favor of either on social values. Crucially, such a choice will not be an indication that they believe either assumption to be true, or even more empirically adequate, than the other. It will be a choice to accept and work with one rather than the other for pragmatic reasons, and it will be a choice that should be revised in light of new evidence that shows the assumption is in fact unlikely to be empirically adequate.

Because empirical adequacy requires only fit with observables, it can be not only that more than one *theory* is empirically adequate for a domain but also that more than one *auxiliary* is empirically adequate for a domain. (Of course, not all auxiliaries involve unobservables, but some do.) When our evidence is incomplete, more than one theory or auxiliary may *appear* to be empirically adequate. In such a circumstance, epistemic considerations would leave us indifferent, in a tie that pragmatic considerations like social values could legitimately break, in a decision that is nevertheless not about which theory or assumption to *believe*. Thus the tiebreaker use of social preferences works just the same for auxiliaries as it does for theories themselves. With Longino, I can say that it can be legitimate to use social values to decide which auxiliary to accept. Moreover, because which auxiliary one accepts decides which evidence is relevant to a hypothesis, this means that social values can have a role in the preconditions for evaluating evidence. However, that role is not *reason to believe* an auxiliary assumption, and on the CE view, it should never be the case that social values are deployed in deciding which statement to accept before our current epistemic resources on that question are exhausted. This is the precise sense of my commitment here to not condoning the entry of social values before, or as a precondition of, epistemic evaluation.

The ideal I describe demands pure epistemic judgments before social values can be legitimately taken into account in judging auxiliaries. Admittedly, there is a sense in which on the CE view the subsequent judgment of whether evidence favors a hypothesis will have involved social values, insofar as social values may have played a role in the choice of the auxiliary that determined which evidence we hold a hypothesis up to. Nevertheless, neither in that choice nor in the subsequent judgment of the hypothesis do social values act as a reason to believe a statement: The subsequent epistemic judgment is whether *given* the chosen auxiliary, the hypothesis is confirmed by the thereby chosen evidence. Thus, the view allows that social values have a legitimate role in theory choice, both in the choice of auxiliaries and in the choice of hypothesis, but at every stage it demands pure epistemic judgments first. The priority of pure epistemic judgments is both conceptual and temporal, because our epistemic resources at a given time must be exhausted for the limits on our pragmatic options to be determinate.

Longino rejects this way of saving her view from the Haack objection, for two reasons. One is that she is not sympathetic to antirealism, a doctrine that, like other controversial aspects of CE, it is not within the scope of this chapter to defend. The other reason is disagreement with

the demand that epistemic and social considerations be strictly separated, that is, that epistemic judgments about auxiliaries (or anything
else) be pure, and that only after we have exhausted our resources for
them at a given time should social values have a role in our choice of
hypothesis. The view I am describing does not accord a place to a claim
that it is, in part, the very point of Longino's view to make sense of,
namely, that it is unrealistic to expect epistemic judgments leading to
theory choice to be devoid of social values. Thus, there is a sense in
which the way that I have defended the overall Longino view, making it
palatable via CE, is giving up her most important claim.

This claim needs to be taken seriously, I grant, at least for the human sciences, because it is a response to the way these sciences actually
behave, and I think to a recognition that there are limits to the capacity
of an individual epistemic subject to be conscious of, much less actively
engaged in scrutinizing, the assumptions that come to him or her in
virtue of participation in a given society, at a given place in that society.
However, I differ from Longino in my response to scientists' failures to
actually maintain clear distinctions between evidence and social values
as reasons to accept hypotheses. First, as mentioned earlier, if we are
constructive empiricists, then a demand that we "directly" confirm assumptions that we make epistemic commitments to is more realistic
than that same demand is for realists, because we make *epistemic* commitments only to claims about observables. At least, the demand in question
is more realistic according to the constructive empiricist. This constructive empiricist response to worries about our capacity to confirm theories
directly (i.e., without relying on social values) differs from Longino's in
that it opts to restrict what we take as the epistemic goal of science to
fewer truths, rather than alter or weaken any epistemic standard for
achieving truths.

Second, while Longino is right that effective scrutiny of auxiliary
assumptions requires more than that the individual think hard, have
good intentions, and follow the best methodological rules, I do not
think this claim is incompatible with maintaining the highest epistemic
demands on individuals. Because the values that seep into our choices
of auxiliaries can be social, their scrutiny requires that the scientific
community meet a number of criteria that ensure that genuine criticism
is happening at the level of social units, that is, communities. To have
objective scrutiny of community-level social values requires interactions

between communities that qualify as critical, and so on. Let me call this the *socialist* view of the regulation of epistemic choices. I will call the view that is implicit in the constructive empiricist's demand that epistemic reasons and commitments be kept distinct from pragmatic reasons and commitments the *individualist* view because reasons and commitments are had by individuals in the first place.

Political connotations of these terms notwithstanding, I regard the individualist and the socialist views as compatible and complementary, because the socialist view does not undermine the idea that epistemic purity should be demanded of individuals, and the individualist demand does not rule out the appropriateness of regulation at the social level. One may worry that accepting the socialist point does undermine the individualist ideal because making demands on the social aspects of a community appears to be motivated by an acknowledgment that the individual cannot be epistemically pure. Relatedly, one might think that if people actually followed the individualist ideal, then the socialist ideal would be otiose. It is not clear to me that the second claim is true because there are ways in which legitimacy can be conferred on a theory through actions or events that no individual person could be held accountable for; think of the verdicts of committees and funding agencies.

What of the first worry, that the socialist ideal undermines the individualist ideal? An easy way out of the bind is to regard ideals we want to impose on the social functioning of a scientific community as, in part, rules of application of the individualist ideal. The motivation for this would be the idea that the individual needs help, and can be helped, in achieving the individualist ideal, and this help is supplied by incentive structures at the social level. This scheme would have us regard the imposition of social standards as, at least, a method for achieving the individualist ideal, something we must do in order to improve our chances of achieving the individualist ideal. On this view, the legitimacy of the individualist ideal and an expectation of achieving it, far from being incompatible with the imposition of social standards, is one of the reasons for that imposition.

If the second claim, that proper behavior at the individual level would render social regulation otiose, is not true, then that provides a further, independent reason to call for social ideals in addition to individual ideals. The idea here would be that we can imagine every individual's behavior as being above reproach and yet the results of research turning out prejudiced because of effects that can perhaps be regulated only at the level of the larger group. For example, it is hardly an actionable offense for an individual male committee member on a given occasion to ignore an individual female committee member when she speaks. We all daydream once in a while. However, if all male committee members ignored all female committee members whenever they spoke, then I think it is fair to say we would have a potentially prejudicing phenomenon that might require action at the level of the group.

To avoid apparent tensions between individual and social ideals, one might also further refine individual ideals. For example, the requirement that an individual's epistemic judgments be pure might be refined to the weaker requirement that individuals have a disposition to regard their views as illegitimate upon becoming aware that they have confused epistemic and pragmatic reasons in them, as well as a disposition to become so aware. This ideal would arguably be achievable by an individual, but its achievement would not make social ideals otiose, because what an individual can become aware of is quite limited. In general, it seems to me that far from social and individual ideals being incompatible or mutually undermining, individuals actually need favorable social structures to have any hope of achieving individual ideals.

However the details may come out, it seems to me that a high probability that people will not in fact attain a (certain formulation of) an ideal is not by itself reason to give up the ideal. And an ideal is especially worth keeping if what people do achieve by means of aspiring to it is worthwhile. What we achieve by means of what I have called the individualist ideal is admittedly a topic of disagreement. Some feminists have counseled rejection of this ideal for what I think is the following reason: The existence of this ideal lulls people into believing they have achieved it when they are aware of no blatant prejudice, and this makes the ideal actively detrimental because prejudices that have a social life are often invisible.

This concern is quite legitimate, but I think that the best response to this concern is to strengthen awareness of the social aspects of a scientific community that make it critical or prejudiced, and relatedly to educate people about the fact that socially conditioned prejudice will not necessarily be obvious and may not be visible even to the well-meaning individual. So, I do not think that this concern gives us reason to give up an ideal that demands epistemic purity of the individual. The reason not to give up the ideal is clear: There are plenty of prejudices that are visible or easily accessible to an individual's consciousness, and there is no excuse for not demanding that each of us contend responsibly with those, and keep our eyes open to others.

8.3 Objections to the Constructive Empiricist Solution

The CE conception I have described as a response to Haack's objection may sound fine in abstraction, but we may wonder whether it fits the way we actually see social values at work in science. Given how difficult it is to find even one empirically adequate theory for a domain, how often are we faced with two such theories between which pragmatic factors that we can recognize as the social values authors like Longino have lately called attention to will have the opportunity to be tiebreakers? First, the need for a tiebreaker does not depend on two theories actually *being* empirically adequate but rather on our having some evidence that they are and (in the best case) no evidence that they are not. Second, the CE-based view I am articulating does not limit the situations in which we may appeal to pragmatic factors to cases of tiebreaking, a topic I will postpone to the last part of this section.

For now, note that apparent ties do happen in just the domains that have been discussed as most susceptible to the influence of social values. One of the most discussed cases of the intrusion of social values into theory choice at the time of Longino's book involved the "man-the-hunter" and "woman-the-gatherer" hypotheses of how the most distinctively human traits evolved in our species. To the question both perspectives regarded as pivotal about how the use of tools developed in the species, the hypotheses each have a plausible answer. On one account, male hunting provided the conditions under which having tools and cooperation gave an evolutionary advantage, and the advantage afforded by spears provided the reason that the size of the canine teeth could decrease at a certain point in time and allow humans to take advantage of diets requiring more effective molars. On the gynecentric account, the development of tool use was a response to the nutritional stress women faced as abundant forests were replaced by grasslands in which food was further afield, and as the conditions of reproduction changed to include

longer human infancy and dependency. The nutritional stress of females was greater than that of males because females fed their young
through pregnancy, lactation, and beyond. Tool use on this account developed much earlier than the stone implements used in hunting, as
women fashioned organic materials into objects for digging for, carry-
ing, and preparing foods. As for the changes in human teeth, female
sexual choice of more sociable partners can explain the decline in the
number of males with the most aggressive-looking canines. Thus inge-
nuity began with the women of the species (Longino 1990, 106–8).

Both hypotheses speculated beyond the data we had, but I see no
obstacle to our understanding them as possible accounts of the data we
had. I do not have to be a partisan of the man-the-hunter view to be able
to see that that hypothesis, along with its auxiliaries, will predict much
of the evidence we have found, and I do not have to be a partisan of the
woman-the-gatherer view to admit that, taken with the auxiliaries that
sustain its relation to evidence, this hypothesis fit the observed observables at least as well as (and probably better than) its rival. What we have
in this case is two hypotheses that the evidence at the time suggested
were both empirically adequate. Thus, on the CE account, one might
have legitimately thrown one's commitment behind one or the other of
these two hypotheses on the pragmatic grounds that one or the other was
in line with one's social values. One would not thereby have believed
that the hypothesis preferred was true, and this last step would not have
provided the grounds one had for believing the account to be empirically adequate. But one might have made a legitimate choice to accept
one or the other hypothesis all the same.

I will consider the next two objections together. Although they
come from opposite sides, my answers to the two are related. First, I can
imagine someone worrying that although I have incorporated social values into theory choice, the fact that the choices I allow to be based on
social values come after, and not as integral to, epistemic judgments
and have nothing to do with truth, whether about unobservables or
about observables, trivializes the role of social values in theory choice.
The second objection, which Haack has made to Longino's account,
says that the situation I consider, where two rival hypotheses are running neck and neck as far as empirical adequacy judgments are concerned, is one where we have no right to make a choice at all because
the evidence is insufficient to distinguish the rivals (Haack 1993a, 35).

The first thing to say about the first of these objections is that choices made on the basis of social values with the purpose of further-ing social ends are trivial only if the social ends are trivial. A theory need not correspond with reality for its ideas and stories to have social effects—effects that we may or may not think it is good to promote. Moreover, people may *regard* the pictures and stories that a theory con-tains about unobservable items as true when some part of the scientific community accepts the theory, even if (or, on the account I am sketch-ing, though) that is not the enlightened attitude to take toward an ac-cepted theory, and even if it is not the attitude scientists actually take. It is legitimate, and not trivial, that we weigh the consequences of such takings to be true by the society at large when we make a pragmatic de-cision about whether to accept a theory that passes epistemic muster as far as we know.

The first thing to say about Haack's objection—that we have no right to make a choice at all in case of ties—is that although it would be wellsuited to address someone whose claim entailed that social values could be a reason to think that one of these tied theories was true, it can gain no traction against the position I am describing. Constructive em-piricism grants that choices about whether one theory is closer to the truth or even more empirically adequate than another must be based on evidence and that we have no right to make a choice about such an epistemic matter when the evidence is (roughly) equal for two rivals. However, the choice between two epistemically tied rivals that accord-ing to CE we legitimately make after the evidence is in is not a choice about which theory is closer to the truth, is more empirically adequate, or has more evidence in its favor. It is a choice about which theory serves better our practical goals, including social goals. And I see no rea-son to think that two theories whose evidence makes them equally com-pelling will necessarily be such that which theory we decide to work with makes no difference to social goals, or such that no one could find social reasons to prefer the one to the other. Whether a theory makes much difference to any social goals does depend on its subject matter but surely not in general on how it stands epistemically when compared with its rivals.

It may be that Haack's intention was to submit that when two theo-ries are tied epistemically we have no legitimate grounds to do anything—including promotion of social goals—with either one. However, this

claim would be far harder to defend than the claim that social preferences cannot be grounds for epistemic choices, and it is not a claim she
even tries to defend. Surely, if the positive evidence in favor of the empirical adequacy of the theories is sufficiently great, then we have a right
to use either theory on observables to serve our practical ends, where
what is "sufficiently great" will depend on the level of reliability the
practical goals demand. We are well acquainted with the shortcomings
of Ptolemy's model of the heavens, but astronomers use the model to
this day because it is the best tool for many calculations of things around
the solar system. However, even leaving social goals aside, we would have
to reject the claim that we have no right to do anything with epistemically
tied theories because of its long-term consequences for achievement of
the epistemic goals of science, a concern that Longino's call for a more
realistic epistemology has taken very seriously.[6]

This is because two theories that are tied according to our present
evidence may not actually be empirically equivalent, even if our evidence so far suggests that both are empirically adequate. Obviously, the
two theories may coincide in what they say about the things we have in
fact observed, without coinciding in what they say about everything that
is observable. (I am assuming that our evidence comes only from observables.) It can be, further, that we simply do not know whether the
two theories coincide in what they predict about all observables. A general proof of the empirical equivalence of two theories is sometimes possible when the theories are axiomatized and mathematical but not as
readily available with the sorts of hypotheses that tend to have the most
relevance to social values. In such a case, only subsequent development
of the theories would tell us whether they are empirically distinguishable.

Further research may be the only route to finding out not only
whether two hypotheses are empirically distinguishable but also exactly
how each can be linked to observables. Kant thought the question
whether the universe was finite or infinite in space and time was an un-
resolvable antinomy of reason, but now we understand how these questions are linked to claims that are empirically testable. Some seem
ready to believe that nothing that *could be* observed would distinguish
the man-the-hunter from the woman-the-gatherer view of human evolution (Haack 1993b, 35), but this is something we do not know and
should be wary of assuming. I doubt that most of us thought we would be
able to distinguish summer and winter seasonal temperature variations

for periods 34 million years ago, that is, to find observational evidence that could be linked to a cooling trend in the winters that was not present in the summers. (Weather, like behavior, is evanescent after all.) But recently a method has been found to make such distinctions via variations in the ear stones found in a certain species of Gulf Coast fish that survived the mass extinction of that era (Ivany et al. 2000). Human behavior of the past is a matter particularly difficult to link tightly to observational evidence, but cannibalism on the part of some prehistoric native Americans has recently been claimed on the basis of chemical analysis of their preserved feces (Wilford 2000).

In any of the sorts of cases I have described, research on both rival theories will be required if we are to determine what further empirical consequences each has and how to test those predictions. If Haack's claim is that during this period of research everyone should remain agnostic about which, if either, of the two theories is *true*, then CE agrees (though that is because, according to CE, we should always remain agnostic about *that*). However, to demand that no scientist work with either theory would make advance on either of them impossible. And to demand that no scientist work on one theory to the exclusion of the other seems an unwarranted restriction on the division of labor. Further, though some scientists will be able to work on the development of both theories by keeping an open outlook, others will not find this psychologically sustainable. This tendency should not be viewed entirely as a weakness, either, because advocacy and competition can bring discoveries in ways that we should not want to impede. I conclude that it is not impermissible to adopt an attitude of advocacy, or what van Fraassen calls "commitment" or "acceptance" for one theory over another on the basis of social values in cases of epistemic tie.

If we are ever to know whether two theories apparently equally supported for the status of empirically adequate are indeed both empirically adequate or rather empirically distinct, and even what their empirical import is, then the scientific community usually has to do more research, and it will be to the positive good of that research if some people make a commitment to one theory while others accept its rival, more to the good than if no one makes any such commitment. If the commitment is made on grounds of social preference, so be it. The fact that commitments to one or the other of the epistemically tied theories can serve the epistemic interests of science in the long run also addresses the former objection, which worried that my view had trivialized the theory choice that is allowed to be made on the basis of social values. Here we see that not only can such a choice have social consequences but also making such a commitment at all to one or the other theory can serve the epistemic ends of science by motivating more research.

The view I am sketching has the feature that we can acknowledge a legitimate role for social values in theory choice while not admitting that social preferences can be reasons to believe a theory true (or empirically adequate) and that we can do this without rejecting the distinction between facts and values, or between the contexts of discovery and justification, and without denying that theories have truth values. One may wonder, finally, whether the price of these features—features that I regard as advantageous—must be as extreme as some take antirealism to be. In particular, one may suspect that the epistemic tie situation in which I find a tiebreaker role for social values is a situation that the realist could find just as easily in any case where insufficient evidence is (roughly) equal for two rival theories.

There remains a salient distinction between this and the antirealist's tiebreaker situation. The antirealist's tied theories can both be believed to have met all the antirealist's epistemic goals, namely, empirical adequacy (and consistency and empirical strength), because it is possible for more than one theory to do all of these things. For the antirealist, an evidence tie can be an epistemic tie in the sense just described. In contrast, we can know a priori that it is *not* possible for the two theories to have met all of the realist's epistemic goals, because no more than one theory can be true. For the realist, an evidentiary tie cannot be an epistemic tie in the sense I described for any theories that go beyond the observables, because such theories will make reference to unobservables, in the interesting cases they will differ in what they claim about unobservables, and which of two such theories is true is what it is the realist's aim to find out.

The realist's aim is to believe the true theory, and it could not be right to believe true both of two distinct theories of the same subject matter. Thus, if we appealed to pragmatic factors as a tiebreaker in such a case, we would be illegitimately substituting pragmatic factors for missing evidence. Because the antirealist's epistemic goals are more modest, it is possible for a full epistemic tie between two rival theories to arise, both in the sense that both theories in fact meet all epistemic goals and, even more commonly, in the sense that we could have reason to *believe* that both meet our epistemic goals. This is why appeal to other kinds of reasons for theory choice is permissible for the antirealist.

One may object to the steps I have just outlined: If we accept van Fraassen's permissivist conception of rationality, according to which rationality rarely compels us to choose uniquely among possible beliefs and we are permitted to believe anything we are not rationally compelled to disbelieve, then why must we not admit that the realist can believe whichever of the two tied theories she likes better? (van Fraassen 1989, 171ff.). Even though the realist knows both theories cannot be true, that does not mean *neither* can. Why is she not permitted to pick one, just like the antirealist? I think that we can be permissivists about rationality and still come to my conclusion that in the situations I have described, the constructive empiricist is in a position to appeal to social values for theory choice, whereas the realist is not.

The issue for the claim I am making is not whether the realist is permitted to believe one of the tied theories. On the permissivist account, he is allowed, but what matters to my point is *on what basis* the realist or antirealist may believe or accept one of the two theories. Permissiveness about rationality will allow the realist to believe one of those two theories but only on the basis that nothing rationally compels him to reject it. This does not change the realist's epistemic goal, a goal that makes evidence (broadly construed) the only sort of positive reason to believe a theory. One might put it this way: Permissiveness about rationality allows one to believe *without* a reason but does not change things in such a way that the realist may believe a theory citing his social values as the reason. This is because for the realist the choice of a theory is not a choice of which theory to accept but a choice of which theory to believe true. This is a purely epistemic goal, and pragmatic grounds for such a choice are inappropriate.

Since in the case I have described, the realist definitely still has epistemic questions to address—those two theories cannot both be true, and truth is the epistemic goal—other sorts of positive reasons cannot decide between them because the realist has not discharged her epistemic duty. For the antirealist, when the two theories are tied, that does not mean she knows that both *are* empirically adequate; for any interesting theory, there is always evidence still out. Nevertheless, it is not irrational to believe that they are both empirically adequate on the basis of evidence suggesting that, because it is possible that they are. (Here I not only acknowledge but also rely on permissiveness about rationality.[7]) And once one believes this, other sorts of reasons, pragmatic reasons, become legitimate positive reasons for choosing which theory to accept. The realist, by contrast, can never get to the point where other reasons become legitimate positive reasons for acceptance. This is because if the realist ever got to the point where she had discharged her epistemic duty, there would be only one theory left to choose from.

What I should really say is that the realist can never get to this point where nonepistemic factors become positive reasons for theory choice *unless that is where he started*. I am assuming that both the constructive empiricist and the realist are committed to the priority of epistemic goals over pragmatic goals, and therefore epistemic criteria over pragmatic criteria, for theory acceptance. If we assume that both the constructive empiricist and the realist regard epistemic criteria as having to be satisfied before pragmatic criteria can become relevant, which I think is an assumption commonly made on either view when thinking of pure science, then the constructive empiricist can be let off the hook and allowed to consider pragmatic factors as relevant to theory choice, whereas the realist never can. Put differently, the constructive empiricist can allow pragmatic factors to affect theory choice while never giving up the priority of epistemic over pragmatic criteria, whereas the realist cannot.

As I mentioned earlier, the constructive empiricist does not need an epistemic tie between two theories that both appear to be empirically adequate to be allowed to appeal to social values as a reason to accept a theory. Let us consider the worst possible case: The theory he favors is significantly less well confirmed than some rival; it is an epistemic dark horse. Let us assume, to make things cleaner, that his theory has no clear counterinstances. The darkness of its prospects comes simply from a lack of positive evidence or positive success. What may the realist do? What may the constructive empiricist do?

If we assume the permissivist conception of rationality, the realist may believe the theory is true. What she may not do is believe it is true because of her social values. May her social values ever enter her choice of theories? Arguably not, stemming from the fact that her epistemic goal is to believe true theories, and I am taking epistemic goals to be primary for both the realist and the constructive empiricist. Permissively speaking, the realist may believe the theory is true on slim positive evidence and for no other reason, but her epistemic task can never be believed to be done, except in a circumstance that leaves her with no more choices to make, that is, when she believes one theory to be true. Thus in the circumstance where the realist's epistemic task is believed to be completed (for the moment), she is left with no more choices to make and so no choices to make on the basis of social values.

The constructivist empiricist fares differently for reasons that are already familiar. Assuming permissiveness about rationality, when the constructive empiricist is faced with two theories and one has much more evidence, though neither is certainly falsified, he may believe that the one with less evidence is empirically adequate. Given that the other has more evidence, it would be odd, then, not to grant that it probably is empirically adequate, too. (If there is a reason not to grant that, then the constructive empiricist's epistemic criteria would be enough for him to decide to accept the theory he likes, so it would not be a case of interest.) For the moment, then, his epistemic work is done, but he is still left with a choice because both theories are believed to meet his epistemic goals. So, he may choose which theory to accept on the basis of his social preferences.

To recap, there is a way of accommodating the influence of social values on theory choice without falling into the fallacy Haack warns against, an option that is open to CE but not to realism. Is there any other strategy for achieving this end that a realist could accept? Of course, I do not know all the strategies that may be open to a realist, but I will close by discussing why I think one option that looks attractive is not successful.

In this option, we accommodate Longino's claim that social values are ineliminable from theory choice in science by granting that the role of social values is significant but maintaining that it should be, and largely is, played out entirely within the context of discovery. Such values will be among the things that lead people to articulate certain theories rather than others, but rigorous testing devoid of social values is the only way a choice to believe a theory true can be justified. Maintaining in this way that social values have no legitimate role in the context of justification sidesteps the Haack objection. Constructive empiricism has no reason I can see to give up the distinction between contexts of discovery and justification, but the view now under consideration has it that this distinction all by itself can resolve the tension that I described CE as resolving, thus making antirealism unnecessary. There are signs in Longino's text that she is confident that social values are ineliminable from the choice of auxiliaries only while a theory is under development, and they are encouraging for this proposal. After theories are developed, she admits, independent confirmation of auxiliaries is at least sometimes available (Longino 1990, 51–52).

This strategy does not sit well with Longino's dim view of the purported distinction between discovery and justification, but I think part of her disapproval can be addressed. It is true, as she points out, that the context of discovery has often been defined in terms of the psychological histories of individuals; through mysterious processes involving dreams, guesses, and other parts of mental life, people light on novel ideas (Longino 1990, 64–65). This is why the distinction is often assumed to run parallel to the distinction between psychology and logic. It tends to be assumed that the mental processes leading to discovery are randomizing because various and unknown. Longino is fully justified, I think, in complaining about this conception of the distinction, because to define it so is implicitly to deny that which ideas and theories come into existence tends to be systematically related to the culture, social structure, or socioeconomic interests of the context in which an individual scientist works. This is to deny something that is true, in the human sciences at least.

However, the context of discovery can be defined in more careful terms: This is the context in which the *genesis* of ideas and theories takes place, the context composed of the causal or temporal history of the genesis of ideas and theories. The genesis of ideas has many aspects, including social ones; all aspects count. This account of the context of discovery allows us to reject as we should the presumption that the ideas and theories generated represent anything like a random sample of the possible ideas and theories. With this understanding of the distinction in mind, the proposal under consideration is this: that we regard the use of auxiliaries that have not yet been "directly" confirmed as something that takes place in the context of discovery. This means we require that auxiliaries be "directly" confirmed *at some point*, and we keep in mind that an auxiliary is not epistemically justified, and belief in any theory it is necessary to the confirmation of is not justified, until the auxiliary has been "directly" confirmed.

If Longino is right in her descriptive claims, then a theory's auxiliaries often do not get "directly" confirmed until after at least the theory's development. Those auxiliaries may not get scrutinized or justified, and hence the theory is not fully scrutinized or justified, until some time after the theory has taken on a certain life with the auxiliaries' as yet unjustified help. This suggests an alternative diagnosis for cases in which social values have affected theory choice in science: Auxiliaries get used in the context of discovery (i.e., without "direct" confirmation) for so long and with such robust effect on the development of a theory that the attention a theory has received creates a presumption in the theory's favor that is mistaken for epistemic warrant. This suggests that the best prescription for science is not to abandon the requirement that epistemic judgments be pure and primary but to reiterate the requirement that a theory not be considered justified until after its auxiliaries are "directly" confirmed. What Longino has observed, on this view, is people overstating their cases for their theories, a phenomenon we should speak out against and not condone.

This warning not to regard a theory as justified until its auxiliaries are "directly" confirmed, regardless of the attention it has received, is reminiscent of Haack's insistence that when "available evidence is not sufficient to decide between rival theories . . . [t]he proper response is that, unless and until more evidence is available, scientists had better suspend judgment" (Haack 1993a, 35). However, the strategy just described that Haack's stricture here seems to be allied with fails to resolve the problem at hand: If we refrain from regarding a theory as justified, it does not follow, as this view needs to assume it does, that everything that happens concerning that theory is rightly assigned to the context of discovery. First, there obviously exist questions of legitimacy about a theory, even when we are not at the point of choosing whether to believe that it is thoroughly epistemically justified. Second, we give answers, even positive answers, to questions of justification, even when the answers are not full-blown belief that a theory is true, and we are obligated to give these provisional answers to evaluative questions about a theory's prospects as we understand them. The context of justification is not the context merely of the justified.

This is because development of theories is not blind to justificatory questions about the theories' worth. It could not afford to be because development requires resources, and resources must be differentially allocated due to their scarcity. Moreover, judgments about allocation of resources for development of theories often have to be made on the basis of slim positive evidence; following the recommendation to suspend all judgment in such cases would close down science, and refraining from regarding a theory as fully justified in such a case is good but irrelevant to the evaluative questions we do need to answer. (We can even imagine a case where a judgment is necessary despite the slim evidence for two theories being tied, simply by imagining that developing either at all would require more than half of the available resources.) It is thus no surprise that explicit arguments about the prospects and worth of a project are demanded of scientists who seek funding even for development of a theory. The winning theories are judged to be to some degree, in some way, more justified than their competitors, and their consequent ability to be developed due to the funding won is an indication of that judgment.

It is wrong to think that there is no justificatory question that we ask or answer except the question whether a theory is fully justified and thus mistaken to think that development of theories is an activity that can be assigned completely or even primarily to the context of discovery. The context of discovery is a place where things just happen. Development of theories is not something that just happens. To think so looks like a way of sweeping under the rug decisions that require justification. Thus, the strategy open to the realist of acknowledging social values by confining them to the context of discovery is unsuccessful. Constructive empiricism, by contrast, allows social values to be answers to justificatory questions about which theory to choose. The questions are not epistemic, but they do not have to be.

NOTES

1. Constructive empiricism is a restricted form of realism because it grants that theories have truth values and takes the aim of science to be finding theories that fit the truth about observables. I have called it antirealism in recognition of the fact that many realists wouldn't regard this much of their doctrine as worthy of the name.

2. I say there is "at most" one true theory to accommodate the possibility that the true view cannot be expressed in what we regard as a theory. If the truth about observables could not be expressed in a theory, then there would also not be many theories that are empirically adequate, which is why I say there are "potentially" many such theories.

3. I am using the terms *hypothesis* and *theory* interchangeably.

4. "Direct confirmation" in Longino's usage does not appear to require independence from all theory, only independence from the influence of social values and freedom from the regress that threatens once we take the role of auxiliary assumptions into account.

5. Discussion at conference "Value-Free Science: Ideal or Illusion?" in Birmingham, Alabama, February 2001.

6. Though this will not be the basis of my argument, note that a theory is always epistemically tied with its notational variants, but that does not mean we refrain from using any of these theories.

7. I seem to rely on a strong dose of it: One might object that we know a priori that the probability of our coming up with *two* empirically adequate theories for a domain is nearly zero. We may think this, but it is too speculative to be compelling. Given permissiveness, it is enough if the antirealist does not have compelling reasons to think the two theories in question are empirically inequivalent, which he often will not when they are under development or nonmathematical.

REFERENCES

Haack, S. 1993a. "Epistemological Reflections of an Old Feminist." *Reason Papers*, 18, pp. 31–44.

Haack, S. 1993b. "Knowledge and Propaganda: Reflections of an Old Feminist." *Partisan Review*, 60, pp. 556–63.

Ivany, L. C., et al. 2000. "Cooler Winters as a Possible Cause of Mass Extinctions at the Eocene/Oligocene Boundary." *Nature*, 407, pp. 887–90.

Longino, H. E. 1990. *Science as Social Knowledge: Values and Objectivity in Scientific Inquiry*. Princeton, NJ: Princeton University Press.

Longino, H. E. 1994. "The Fate of Knowledge in Social Theories of Science." In F. Schmitt, ed., *Socializing Epistemology: The Social Dimensions of Knowledge*, pp. 135–57. Lanham, MD: Rowman and Littlefield.

van Fraassen, B. C. 1980. *The Scientific Image*. Oxford: Clarendon.

van Fraassen, B. C. 1989, *Laws and Symmetry*. Oxford: Clarendon.

Wilford, J. N. 2000. "New Data Suggests Some Cannibalism by Ancient Indians." *New York Times*, September 7, 2000.

NINE

THE VALUE LADENNESS OF SCIENTIFIC KNOWLEDGE

Gerald Doppelt

9.1 Introduction: What Is at Stake in Value-Free Science?

There are few philosophical distinctions as central to twentieth-century thought as the distinction between fact and value. Indeed, the distinction has proven its utility for enlightenment and emancipation by providing a powerful tool for exposing ideological distortion and political manipulation. When individuals or groups confuse fact with value or value with fact, when they embrace values on the basis of distortions of fact or embrace facts on the basis of distortions of values, and when they are unconscious of the irrational mechanisms at work, the fact-value distinction may be deployed as a powerful resource of critique and enlightenment. Certainly, this logic played a key role in the process through which many came to question the value of racial segregation or gender hierarchy, on the basis of an empirical-scientific critique of the pseudo-science(s) and factual distortions underlying beliefs in essential difference or inferiority.

Yet, in the last decades of the twentieth century, it has become fashionable to claim that facts and values are both socially constructed and depend, for their credibility, on the subjective interests or needs of specific groups—which is what facts and values embody or express, rather than any independent world of nature or morality.

In this chapter, I hope to clarify these issues by a rigorous focus on post-Kuhnian insights and debates concerning the relativity of scientific knowledge and their impact on rationality, objectivity, and realism. From my standpoint, the focus on scientific knowledge will take us to the nub of the question of whether science is, or can be, "value free." The claim that science, as we know it, is not and cannot be value free, while gaining an almost theological status for many scholars today, obscures several distinct dimensions of science and the roles different sorts of values may or may not play in each of these dimensions.

We can and should recognize that human beings' value commitments and interests shape and inform the practices of science(s) in many ways. But many or perhaps all these modes of influence may not imply the value ladenness of scientific knowledge itself. In particular, we can acknowledge that value commitments and practical interests typically shape scientists' motives for practicing science; the particular questions and problems they tackle; the concrete concepts, methods, and theories they embrace; and the uses to which scientific knowledge is put, including who has power over its uses, for whose benefit, and at whose expense. Beyond that, we can also recognize that value commitments and interests inform the direction of scientific funding or patronage more generally, the distribution of credit or recognition for scientific work, the division of labor in science, the institutions that structure scientific work, the education or training of scientists, and the class, racial, ethnic, gender, and religious composition of scientific groups or communities.

If and when we recognize all these ways that groups' interests and values typically shape the practices of science, the whole game seems to be over for many scholars: value-free science is an illusion, end of story! Yet none of these lines or types of normative influence, singly or taken together, necessarily implies that scientific knowledge is itself defined by, or relative to, this or that group's interests or value commitments. All of these value dimensions of science aside, the question of whether a group succeeds in producing scientific knowledge may be simply and primarily a matter of whether its theories accord in the right ways with the relevant domains of empirical evidence. Do they succeed in explaining, predicting, or unifying already well-known patterns of phenomena with accuracy? If they do so succeed, it may be this empirical success *alone* that provides the criterion of scientific knowledge and reality. Value commitments can shape the knowledge we seek; the concepts, methods, or hypotheses at our disposal; our motives; how the "we" is constituted; and the like, but it still may be the case that none of these determines whether it is scientific knowledge that we have achieved, when it is indeed achieved.

There are undoubtedly many scientists, philosophers, and laypeople who accept some version of this picture of "pure" scientific knowledge. For them, when science succeeds, its empirical success alone provides the yardstick of knowledge and the true mirror of nature. To this extent, in its moments of triumph, science is value free. Thus, the bottom line for the defense or rejection of the value freedom of science concerns the not-yet-dead, age-old question of the nature or criterion of (scientific) knowledge. Is knowledge itself value free? Must it be value free to possess the objectivity, rationality, and justification that knowledge seems to require? On the other hand, if scientific knowledge is value laden, can the fact-value distinction nonetheless be revised and preserved in some new form?

9.2 Scientific Knowledge and Epistemic Values

I want to defend the view that scientific knowledge is itself essentially value laden but that the value commitments or interests at stake are, in the final analysis, cognitive or epistemic values. I have three sorts of value commitments in mind: (1) normative commitments to *the value* of certain kinds or patterns of phenomena and *not others*, as what the theories of a science must or should explain, predict, and so on to constitute knowledge of the domain of that science (e.g., a physics of motion); (2) normative commitments to (or interests in) *the value* of certain kinds or types of theory *and not others*, such deterministic or indeterministic, possessing this or that mathematical structure, explanatory and/or predictive, as what theories must or should be like to constitute scientific knowledge of the domain (or in the discipline); and (3) normative commitments to the *value* of *certain* kinds of inference, reasoning, or proof and *not others* (deduction, empiricist induction, the method of hypothesis, the consilience of induction) as how scientific theories must or should be established to constitute a knowledge of phenomena. These three sorts of value commitments are bound to the historical development of science (or natural philosophy) as a tradition in which knowledge makers necessarily share such value commitments up to a point, contest them up to a point, and transform them in great and small scientific revolutions. Scientific knowledge thus requires a community of practitioners who implicitly share a certain normative consensus— commitments to the value of *certain* (but not other) types of phenomena, standards of reasoning, proof or argument, and virtues of theory— as "the" values or standards that must or should be jointly actualized by their practice if genuine knowledge is to be attained. Indeed, among these epistemic values, I also include standards of acceptable empirical approximation—a ubiquitous feature of scientific inquiry. Such standards in a scientific group "tell" its members the degree to which a theory Ts observational implications and predictions O may deviate from the measured values of actual observations and yet still be taken as confirming T. Because scientific groups are rarely in a position to employ theories that generate an exact match between the predicted values of observational variables and their measured values, agreement on standards of approximation are all that stands between confirmation and disconfirmation, success and failure in explaining or predicting "the" observed phenomena.

On my view, a shared commitment to these three sorts of epistemic values does not *only* affect *what* scientists seek to know, the *content* of knowledge. Nor does it *just* affect the aims, means, and methods they employ. Pragmatists, reliabilists, naturalists, instrumentalists, empiricists, and realists can accept this much concerning the value ladenness of inquiry while rejecting the value ladenness of knowledge itself. Pragmatists can allow that the aims of scientific inquiry and theoretical knowledge may alter, without affecting the standards and determinants of knowledge itself. For example, scientists can embrace the value of explaining new sorts of phenomena or give up the aim of explanatory theories in favor of ones that yield more reliable and accurate predictions. The pragmatist, instrumentalist, or naturalist can recognize this change in epistemic values while firmly maintaining that what makes any of these scientists' results genuine knowledge remains unaffected and unitary. In this vein, one can reasonably claim that whether it is explanatory or predictive theories we value, in both cases, theoretical knowledge requires that the same unitary standard of confirmation, or the same cognitive mechanism of truth-conducive reliability, is satisfied. My thesis concerning the value ladenness of scientific knowledge contrasts sharply with all such views. My claim is that in some historical cases, a change in aims, problems, or methods amounts to a change in the criteria of a good, well-confirmed, or true theory. Scientific debates concerning the aim of explanation, as against mere prediction, are thus sometimes debates concerning what virtues a theory must have in order for a relation of confirmation to obtain between a theory and a body of evidence.

To claim that this is "sometimes" what is at stake does not quite do justice to the epistemological force of my claims for philosophy of science. If the development of science involves normative shifts in the very criteria of theoretical knowledge, an adequate philosophy of science will need to account for the rationality, objectivity, and progress such normative change may exhibit. Indeed, this remains the case even if we recognize a significant measure of normative continuity as well. To bring the point home, let me reformulate it by analogy with the more familiar notion that knowledge is always relative to available evidence. Operating with the classical notion of knowledge as true, justified belief, philosophers of science typically treat scientific knowledge as relative to well-established bodies of observational evidence. When the body of evidence changes, what scientific groups know, and are justified in believing, changes. My thesis extends this fairly uncontroversial account of the relativity of knowledge to evidence (and incidentally to belief) to encompass the relativity of knowledge to epistemic values and standards. This is a natural development if and only if we agree that evidence is not just a matter of empirical discovery or growth in observational data, though it is that.

Rather, evidence is also a changing normative notion in the life of science that reshapes the matter of which body of evidence or discoveries constitute the domain of phenomena to which theories in an area of inquiry are properly accountable; what sort of inferential or cognitive relation must obtain between this evidence, E, and a theory, T, before E is rightly taken to confirm the truth of T; and what virtues any T must possess — by way of simplicity, causal structure, conceptual coherence, explanatory virtue, consistency with more fundamental theories, and other matters — for T to be genuinely confirmable as true by E. If knowledge is relative to evidence, and I am right that evidence is relative to these three sorts of standards or values, then knowledge is relative to these epistemic values as well. Of course, relativity does not imply any radical claim of relativism or incommensurability, as I will argue in the last section of this chapter. Provided that there are good reasons for embracing (1) new evidence and (2) new epistemic values that redefine the significance of evidence for theoretical truth, then neither the relativity of knowledge to evidence or to epistemic values justifies a relativistic rejection of cognitive progress in our knowledge of nature.

On the other hand, my value-ladenness thesis would dramatically alter the nature of our account of scientific knowledge and its development. Our account could no longer be confined to the familiar picture of a succession of theories, each far better confirmed than its predecessors by an ever-growing body of empirical discoveries and more accurate observational data, in accordance with timeless standards and values. Rather, a measure of practical rationality comes into the picture: an account of the rationality of scientific groups' choices to embrace epistemic values and standards, in some degree different from those embraced by rivals or predecessors. What happens to the fact-value distinction, scientific objectivity, realism, truth, confirmation, and cognitive progress? Such key notions need not be abandoned. But, as I argue later, our way of understanding these notions would need to be revised to accommodate the place of practical rationality in the attainment of theoretical knowledge in science.

But I am way ahead of my game here. Are there good reasons to accept the view that as knowledge is relative to evidence, it is also relative to epistemic values or standards that determine what counts as relevant "evidence"?

9.3 The Defense of the Value-Ladenness Thesis

What justifies my thesis of the value ladenness of scientific knowledge? It is useful to contrast the argument I favor with that of Helen Longino, who seems to defend a similar position (Longino 1990). Longino begins with the classical arguments concerning the underdetermination of theory by evidence and observation. She uses the fact of underdetermination to show that the link between theory and evidence cannot be that of induction, abduction, or logic alone. Rather, the link is provided by scientific groups' commitment to various background assumptions, some of which may involve wider social values, as well as more internal epistemic values. For Longino, values and social interests fill the gap between theory and evidence that the pure logic of inference opens up. Longino's starting point is an apt one for philosophy of science, given the centrality of logical models of confirmation, prediction, explanation, and so on within twentieth-century Anglo-American positivism and empiricism.

My starting point is not the logic of underdetermination but rather how the history of science bears on the nature of scientific inference and knowledge. My thesis concerning the value ladenness of scientific knowledge builds on a series of papers I have written from 1978 to the present (Doppelt 1978, 1980, 1981, 1983, 1986, 1988, 1990, 2001). This work begins with my interpretation and defense of Kuhn's *The Structure of Scientific Revolution* in 1978, a time when Kuhn had few if any friends among philosophers of science and many detractors (Doppelt 1978). Influential interpreters of Kuhn such as Scheffler and Shapere developed a standard, deflationary reading, on which the key to scientific revolution or "paradigm" change, for Kuhn, is a wholesale change in language, meaning, ontology, and worldview, generating a radical incommensurability between scientific paradigms or theories (Scheffler 1967, 1972; Shapere 1964, 1966, 1971). On my counterreading, the more plausible and epistemologically significant key to scientific revolution, or paradigm change, in Kuhn's work is a normative shift in the epistemic problems, data, standards, and values taken to be required of theories for genuine scientific knowledge (Doppelt 1978). This line of argument in Kuhn does not imply radical incommensurability or wholesale change, but such historical shifts in the standards and goals of theoretical inquiry pose challenges to standard philosophical accounts of scientific rationality, knowledge, and progress, so I argued (Doppelt 1978). I pressed this argument by utilizing it to critically evaluate accounts of scientific rationality and progress in the works of post-Kuhnians such as Laudan and Shapere, as well as others of a naturalist, reliabilist, or empiricist persuasion (Doppelt 1983, 1986, 1988, 1990, 2001). With these arguments in the background, I want to clarify the thesis of the value ladenness of scientific knowledge in its general form and explore the matter of how it may inspire a larger conception of the rationality of scientific argument and advance.

I want to defend my thesis of value ladenness as providing the best explanation of the historical development of science and the kinds of rationality it arguably possesses. Arguments concerning underdetermination are not a suitable starting point for me, because, in the post-Kuhnian environment, little follows from them. In particular, such arguments show that standards of confirmation or evidence in science cannot be captured in purely or exclusively logical terms. Many a realist, reliabilist, naturalist, and instrumentalist now blithely accepts this point without being pushed any closer to the thesis of value ladenness I want to defend. The reason is simple. The argument from underdetermination allows philosophers many options for defending a nonlogical but universally applicable standard, or cognitive mechanism, of theoretical knowledge in science, which is supposed to be free of any irreducible context-bound value ladenness. The contemporary scientific realist's appeal to "inference to the best explanation" and the normative naturalists' appeal to truth-conducive, reliable mechanisms of belief formation provide two examples of such viable options, which I take up later. So, I don't think the defense of value ladenness gets much support from underdetermination in our present context. Rather, its best defense is establishing that it provides a better explanation of the historical development of science, successful science, and scientific debates than its main rivals today. In addition, I will argue that the thesis of value ladenness can be deployed to develop a "critical theory of scientific argument" that promises to make social controversies over what is known more rational, especially when social and behavioral science is at issue. As such, the thesis of value ladenness would also be justified by its normative role in providing a more rational evaluation of knowledge claims in contexts of politically charged epistemological conflict.

The defense of the thesis of value ladenness also depends on one's picture of how normative commitments are supposed to inform the attainment of theoretical knowledge in science. On one view, contextual values exert causal influence over which background assumptions are embraced by scientific groups and thus over the theoretical conclusions they draw from bodies of evidence. On my picture, such causal influence is far less central. Rather, some of the background assumptions that make theoretical knowledge possible function as a scientific group's epistemic values, or standards, governing theory assessment. From my point of view, it matters little that wider values or interests, like many other factors, exert causal influence over these epistemic standards. Indeed, it matters only if we can evaluate such wider values or interests to determine when they do or do not provide good reasons for preserving or changing the epistemic standards that drive scientific inquiry. The reason this matters is because the thesis of the value ladenness of knowledge brings a new issue of scientific rationality to the forefront, namely, what is the structure of the practical rationality that bears on the maintenance, criticism, or transformation of the epistemic values at the heart of scientific knowledge? If scientific knowledge is value laden in my sense, then its justification possesses two dimensions: the justification of theoretical belief, relative to evidence, and the justification of the epistemic standards, relative to which the evidence succeeds in providing good reasons for the relevant theoretical beliefs. In the model I suggest later, wider practical values can justify epistemic standards, but so do various other cognitive considerations.

I have said that on my picture, epistemic values enter into some of the background assumptions scientific groups rely on to frame theoretical hypotheses on the basis of evidence. How can we determine when background assumptions are just empirical beliefs, however fundamental, and when they function as values, prescriptive norms, or standards? There are three kinds of evidence in the behavior of scientific groups to establish that a shared belief, goal, method, or procedure constitutes a fundamental epistemic value or standard in my sense: (1) their choices concerning what sorts of theories to pursue, believe, consider, or dismiss out of hand; (2) their grounds or reasons for these choices; and (3) the principles they appeal to in scientific controversies among members of the group and between them and advocates of rival conceptual and/or procedural approaches. The value-ladenness thesis is confirmed if and only if the attribution of a shared epistemic standard(s) of theory assessment to the group is part of *the best explanation* of (1), (2), and (3).

For example, consider how to explain and evaluate the long transition from alchemy to the new chemistry of Lavoisier and later Dalton. As Kuhn and Shapere have emphasized, many of the chemical effects explained by the new chemistry were either unrecognized or marginalized by the alchemists (Kuhn 1970a, pp. 99–100, 107, 133; Shapere 1984, pp. 126–29). Similarly, the alchemists sought to account for and control many observed phenomena (concerning the sensible qualities of things) that are simply abandoned, though their existence is not denied, by the new chemistry. On standard accounts, these transitions are represented as changes in scientific concepts and beliefs, often accompanied by the assumption that the more modern beliefs were better confirmed by "the" observational evidence than the beliefs they replaced. Thus, the change in chemistry involves different beliefs concerning which phenomena of nature embody lawlike natural kinds and reveal the true natures of the common causes underlying them. And the new chemistry enjoys much greater empirical success with "the" observational evidence than its predecessor(s).

The thesis of value ladenness does not seek to falsify this account. Rather, it provides a richer explanation that is capable of answering questions ignored or begged by the standard account. Using the three sorts of evidence I provided, we may learn that the change in the practice of chemistry involves more than the standard change in belief, plus greater empirical success with "the" observational evidence. We may learn that rival epistemic values were at stake concerning how to define the chemical phenomenon that different groups took to be essential to a genuine knowledge of nature or rival standards concerning what a chemical theory needed to explain, predict, or control to yield a true understanding of the properties of substances (Doppelt 1978). This account can explain losses, as well as gains, in the observational explicanda of science, for example, how and why a new chemistry could grow up and succeed, even though it abandons and fails to explain obvious phenomena (why all metals have metallic qualities in common) at the center of previous chemistry (alchemy) and, in principle, still awaits some theoretical explanation (which is achieved in twentieth-century science). Although this does involve new beliefs and empirical success, it is the normative and prescriptive dimension of epistemic standards that provides a better explanation of why chemists at some points ceased to accord any scientific legitimacy to the empirical problems and phenomena at the center of alchemy. They have abandoned not simply the beliefs, concepts, methods, and goals of alchemy but the whole "value-laden" conception of chemistry underlying these components. Without this dynamic of normative consensus, order, and disorder, we cannot give an adequate account of the production, transmission, transformation, and improvement of scientific knowledge.

Clearly, such normative consensus, while necessary for the attainment of scientific knowledge, is *not* sufficient. Success or failure in the achievement of community values and thus scientific knowledge is obviously contingent on many factors and forces at play, among others (1) the way the world is, as realists argue; (2) how effectively scientific groups are able to renegotiate their common values, standards, and goals when they conflict and fail to be conjointly realizable in their practice of inquiry, as social constructivists argue; and (3) the ability of scientific groups to develop theories, techniques, procedures, devices, and the like that provide empirical success and meet their standards, as empiricists stress.

If I am right, socially shared and enforced value commitments are essential to scientific knowledge. Moreover, fundamental scientific controversies and large changes in the way science is practiced typically involve conflicts over, and transformations of, the value commitments that dominate genuine knowledge production. The best way to provide further defense of my view is to consider some of the main objections that can be made against it and some scientific examples that may be taken to confirm it.

9.4 Postmodernism: It's Politics All the Way Down

First of all, postmodern scholars of the politics of knowledge may object that my notion of epistemic or cognitive values depends on a false separation between the epistemic and the wider social or practical interests that define the cultural contexts of all knowledge production. On my view, epistemic and cognitive values are shared interpersonal standards and define the cultural dimension and politics of specific practices or communities of knowledge production. Furthermore, I readily admit that practitioners have reasons or motivations for their commitments to epistemic values that are rooted in wider interests, goals, or values of the society, epoch, and social-gender-racial situation they occupy. Nonetheless, it is the epistemic value commitments I have previously characterized that define scientific knowledge and the normative agreements that enable knowledge-making practices to exist and progress. Put differently, only when such wider social or practical interests get expressed and embodied by special groups of knowledge makers in shared, epistemic, inquiry-defining value commitments, of the three sorts I discussed before, does scientific knowledge become possible.

For example, it is undeniable that the development of modern meteorology by Vilhelm Bjerkness and his collaborators in the first quarter of the twentieth century was driven in part by powerful practical interests in gaining more reliable weather forecasts for aviators, fishermen, and farmers in the context of commercial, military, and political goals (Friedman 1989). The emergence of such practical interests—especially with the age of flight (airships, aviation, etc.)—justified a redefinition of the domain of weather phenomena by Bjerkness to include atmospheric motions and conditions. This redefinition of the weather provided a key epistemic standard or value of "empirical success" for the new meteorology of the Bergen school and its quest for a physics of the atmosphere. A social interest in forecasting certain phenomena of "weather" provided both motivation and a good practical reason for embracing the new epistemic standards for meteorological knowledge concerning what a science of "the weather" needed to include, explain, and predict, according to the Bergen school. Only at the point where wider social values are effectively embodied in such epistemic standards of cognitive success does the possibility of scientific knowledge exist.

Beyond that, the epistemic values that inform knowledge-making practices may persist or undergo rational criticism and modification somewhat independently of those specific wider social and practical interests that initially motivated the credibility of these epistemic values. For example, consider the range of practical aspirations that have motivated astronomers to identify and understand the positions and movements of heavenly bodies. Long after astronomers have given up any hope of reading the heavens to discern the will of God, the outcome of human endeavors, and the like, the epistemic value of certain astronomical phenomena remains central to various branches of scientific knowledge. So although the practices of science are often shaped by a wider context of practical interests and values, the value ladenness of knowledge cannot be reduced to these wider interests, even when they provide good reasons for embracing or revising epistemic values.

This has important implications for an effective normative critique of knowledge-making practices, whether by lay actors or scientific-technical professionals. Effective normative critique demands a challenge to the way established practitioners define the boundaries of the knowledge they produce—the epistemic values or standards embodied in their prac-tice. Those who advance critique may justify their allegiance to rival epistemic values by appeal to wider ethical or political concerns, but such concerns need to be articulated at the level of epistemic values, and transformations in the sorts of phenomena, methodologies, or theories, taken as normative for knowledge making. In addition, rational critique requires some evidence that rival epistemic values meet with empirical success and generate credible knowledge claims in their own right. In the absence of this epistemological dimension, normative critique easily becomes an ideological conflict of interest without cognitive substance. This concludes my reply to the postmodern political critique of science. I will further clarify my argument later in relation to issues of rationality, objectivity, and relativism.

A second set of objections to my view is inspired by philosophers of science who deny that scientific knowledge, in the final analysis, is rela-tive to local, contingent, and historically variable epistemic values and standards, as I maintain. I will consider three such objections to the value-ladenness and value relativity of scientific knowledge (or, for short, the value-relativity thesis), as follows:

1. Against the value-relativity thesis, there are fairly neutral, external, and universal epistemic values and standards in science that provide the criteria of scientific knowledge independent of local context.
2. Against the value-relativity thesis, we can embrace a normative natu-ralism or an externalist reliabilism, which marshals empirical evi-dence for evaluating the reliability or effectiveness of local values and methods in attaining the aims of science.
3. The value-relativity thesis should be rejected because it is incompatible with the evident rationality, objectivity, progress, and empirical suc-cess of science.

9.5 Universal Values and Standards

For the sake of clarity, let me acknowledge the existence of some subset of universal epistemic values in all scientific inquiry: empirical success, predictive accuracy, breadth of explanatory scope and unification, sim-plicity, problem-solving effectiveness, and theoretical truth concerning unobservable causes, to take some well-known candidates. At this level of abstraction, such values could plausibly be taken as motivating at least a great deal of scientific inquiry. This is an important point because agreement on such epistemic values, or a subset of them, provides entire traditions of scientific inquiry across centuries with whatever unity (or continuity) they possess and serves to distinguish them from one another and from other areas of human knowledge and activity. Regardless, such values can function only as criteria of scientific knowledge when they are given flesh and articulated in terms of the more local values and standards to which communities of scientific practitioners are actually committed. Predictive accuracy cannot function as a mark of scientific knowledge in the absence of standards of acceptable empirical approximation. Breadth of explanatory scope and unification do not function as virtues of theory in the absence of standards that indicate which domains of phenomena, or laws, ought or ought not be unified and whether or not unification is taken to require common explanatory and causal mechanism across domains or only common mathematical and formal principles lacking explanatory force (Doppelt 1981, 1986).

Consider the epistemic value of empirical success or the ability of theories to "save the phenomena"—perhaps the value or standard with the strongest claim to scientific universality. Clearly, this goal or norm cannot function as a criterion of knowledge until questions like the following are answered: What sorts of phenomena are most important to "save," and which can be neglected? What kind of theory is valuable or useless to "save" the phenomena? What type of reasoning or proof is valuable, invaluable, or of little value, if the phenomena are to be saved by a theory or empirical law? If a theory saves "the" phenomena, does this provide good reason for taking the theory to be true? Does saving the phenomena require that a theory provide a good or the best explanation of the phenomena, and what are the criteria of a good explanation (simplicity, unification, novel predictions, indirect theoretical support, causal mechanisms, or which combinations of these)? Similar problems arise for Laudan's bold attempts to defend universal standards such as "problem-solving effectiveness" or maximal internal consistency (and mutual realizability) among scientists' aims, methods, and theories, as I have argued elsewhere (Morrison 2000).

In the history of science, groups and communities of practitioners of inquiry answer such questions in quite different ways, linking the very possibility of scientific knowledge to the epistemic values to which such communities are actually committed. What follows are some historical examples of such disagreements over epistemic values.

The first concerns how the domain of phenomena proper to a given science is defined and bounded; for example, many of the observational phenomena thought essential for a chemical theory to explain, on the standards of the premodern chemistry of neo-Aristotelians and alchemists, are excluded from the domain and replaced in epistemic importance by other sorts of phenomena on the standards of Daltonian chemistry (Doppelt 1978; Shapere 1984).

The second concerns the relative epistemic weight, evidential or probative power, or explanatory importance of one as against another kind of phenomenon, evidence, or empirical achievement; for example, phenomena deducible from a theory that are surprising, previously unknown, or different in kind from those that the theory is designed to explain have special or even unique evidential force in the standards of Herschel and Whewell, which they completely lack on the standards of Mill and others (Laudan 1981).

The third concerns the form(s) of inference that must exist between observational evidence and hypothesis if the latter is to gain rational credibility from the former; for example, for the Newtonians, a hypothesis is only knowable on the basis of evidence if it is a strict inductive generalization from that evidence, whereas on the method of hypothesis endorsed by the ether theorists of the day, hypotheses may also be indirectly known on the basis of evidence that implies or explains but does not inductively generalize (Laudan 1981).

The fourth concerns what sorts of hypotheses and entities may gain scientific credibility from the evidence or observational phenomena they imply, for example, Newtonian as against the seventeenth- and eighteenth-century ether theorists concerning the legitimacy of unobservable entities (Laudan 1981).

My argument is that the existence of "universal" epistemic values such as empirical success, explanatory breadth, simplicity, or problem-solving effectiveness does not by itself refute the thesis of the value ladenness of scientific knowledge. Of course, a critic may reply that I have not established the impossibility of formulating a universal value or standard of empirical success (explanation, predictive accuracy, etc.) that is independent of history or context. Only then can we gain a genuinely universal and normative criterion of knowledge, one based on a philosophical analysis or theory of the very nature of science, not a description of what this or that scientific group actually takes knowledge to be.

Of course, this is the traditional method of a priori epistemology that has been effectively challenged by the historicist, sociologist, and especially naturalist turns of our present period. A priori philosophical criteria of scientific explanation, confirmation, verification, and the like always elicit a host of contemporary or historical counterexamples, that is, examples of scientific explanation, reasoning, and the like, which *we value as such* but fail to fit the a priori philosophical model in question. Even if we could come up with a standard that fits much *contemporary science*, how could we justify it as the right and proper basis for all theory assessments, for all time? Rather, the relevant points would be different: Contemporary scientific communities are normatively committed to it, it has been successfully employed to produce explanations, discoveries, or predictions unavailable to previous scientific communities, and the like. For my part, I do not draw the conclusion that philosophy cannot say anything important and distinctive concerning scientific inquiry as a whole. Indeed, I take my claim concerning the value ladenness, or value relativity, of scientific knowledge to represent a distinctively philosophical conception of scientific inquiry through one informed by history and the social sciences.

9.6 Normative Naturalism and Externalist Reliabilism

Among most philosophers today, the rejection of traditional a priori epistemology leads not to my thesis of value ladenness but to science-based or naturalistic epistemology. Here, the project is to utilize the methods of science to characterize scientific knowledge. Normative naturalists can recognize the value ladenness of scientific practices while resisting the conclusion that knowledge itself is value relative. They propose that we evaluate the value-laden practices in science empirically, as more or less effective means to the ultimate aim(s) of science (Laudan 1987). Scientific knowledge can be neutrally or scientifically characterized as those locally valued methods, standards, aims, theories, and so forth that prove in fact to be more effective and reliable than rivals in achieving the aims of science.

Normative naturalism comes in different versions, depending on how the aims of science are characterized. On one view (the multiplicity view), scientific groups even within the same discipline(s) and area(s) of inquiry, may embrace different ultimate aims, for example, prediction rather than explanation, or probabilistic explanation rather than nomological explanation. As I read it, the multiplicity version of normative naturalism concedes most of the value-ladenness thesis. If it evaluates the efficacy of local values and methods relative to aims and allows aims to vary from one scientific group to another (in history or contemporary science), then, on its own view, scientific knowledge will be relative to the larger epistemic values and aims to which some but not all scientific groups are committed. This conforms to my view of value ladenness (Doppelt 1990).

On a second version of naturalism, there is but one unitary aim of all science, for example, to discover the truth about nature. Suitably refined, this view challenges the value-ladenness thesis. We should refine the view to disambiguate realist from instrumentalist goals, though it is not clear how the naturalist or reliabilist will adjudicate the disagreements concerning "the" goal of science among realists, instrumentalists, pragmatists, unificationists, empiricists, and others. For the sake of my argument, suppose we are realists and fix the aim of science as the attainment of true theories, or theories that are maximally reliable for generating theoretical truths concerning the unobservable causes of observed phenomena. Then the naturalist can characterize knowledge as whatever local epistemic values, methods, theories, and the like prove, in fact, to be the most effective means to this aim (Goldman 1986).

The normative naturalist's language of unitary aim, efficiency, and empirical evidence promises a value-neutral science of scientific knowledge. In my view, this language is deceptive. It masks the fact that the attainment of theoretical truth (in science) is no more a value-neutral unitary aim than is empirical success, or saving the phenomena, as I previously argued. Suppose, in the naturalists' fashion, we set out to determine which of the value-laden practices of scientific groups is, in fact, most effective or reliable in producing true theories concerning the underlying causes of empirical phenomena. What epistemic standards must be satisfied by true theories? For example, should true theories be explanatory or predictive, or simple and unifying, or deterministic, in each case—as some, but not all, scientific communities or groups have insisted or denied? Do we value and count as true any theory that succeeds in accounting for or implying already well-known kinds of phenomena, as Mill and other scientists advised? Or do we restrict the value of theories, and thus those that we will count as true, to theories that succeed in predicting previously unknown, surprising phenomena, different in kind from those they were designed to explain, as Whewell and others advised (Doppelt 1990, pp. 11–14)? And which sorts of phenomena define the most central, valuable, or significant core of the domain, the phenomena it is most important for a true (or reliable) theory to be true to or of? The normative naturalist seeks to circumvent all of these issues, either obscuring the questions concerning the epistemic values they raise or smuggling in the epistemic values of current scientific communities and practices.

Reliabilist epistemologists embrace an externalist standpoint that promises to make knowledge independent of the reasons, evidence, standards, and epistemic values internal to the knower. Knowledge is simply a matter of whether the knower is using a reliable or truth-conducive mechanism of forming beliefs. Knowers need not know that they are using a reliable, truth-conducive cognitive mechanism in order to know. Of course, the same would be true for scientific communities whose knowledge, for the externalist and naturalist, depends not on their own internal standards and values but simply on the reliability of the mechanisms or methods by which their beliefs are formal.

But surely, someone has to know which cognitive methods and mechanisms are reliable and truth conducive, if externalist epistemology is to reach any conclusions about knowledge. Presumably, the naturalist epistemologist possesses a scientific knowledge of which mechanisms are reliable and truth conducive. But can the naturalist gain a scientific knowledge of reliability without a commitment to some of the very same particular epistemic standards at stake in the value-ladenness thesis? Which mechanisms of cognition, if any, are reliable in generating theoretical truths about unobservables? How is the naturalist supposed to adjudicate the normative dispute between scientific realists and instrumentalists or empiricists? They disagree over whether inference to the best explanation is ever a reliable method or mechanism for arriving at theoretical truths and, more generally, over the question of whether any mechanism produces theoretical truths in science.

How can the naturalist know whose or which mechanisms are reliable and truth conducive without a commitment to a realist or antirealist view of whether theoretical truth in science is ever possible? Furthermore, if a scientific group embraces standards on which theories count as true only if they are consilient, simple, explanatory, predictive of novel phenomena, or deterministic, why should its knowledge be determined by reliabilists, if they have lower or different standards of theoretical truth in science, than that of the practitioners themselves? Reliabilism is as value laden as the bodies of scientific knowledge it hopes to evaluate "externally" and "naturalistically." Given that this is a case of one group with scientific pretensions (naturalists) evaluating the knowledge claims of another, whose epistemic standards ought to be defining truth, truth conduciveness, and reliability? The externalist enjoys no epistemological privilege in determining whose standards of truth and reliability define scientific knowledge. The influential idea that naturalism, reliabilism, and externalism provide an escape hatch from the Kuhnian world of shifting epistemic standards is a scientistic illusion: It posits a standpoint of empirical neutrality—the myth of "reliability" as the gold standard—that is either empty or as value laden as the knowledge claims it seeks to evaluate, "externally."

Put differently, normative naturalism requires an epistemology, some account of the standards of reliability or confirmation that the naturalistic scientist of scientific knowledge can use to determine which methods, goals, and theories are the most truth conducive and thus warranted as scientific knowledge. But short of producing and justifying some such universal standard (who has done that?), the naturalists' criterion raises all of the same issues of value relativity that "empirical success" raises. On the other hand, once a scientific community or tradition has implicitly committed itself to the value of predicting and explaining certain sorts of phenomena, the value of certain standards of reasoning, and the value of certain virtues of theory (or models), then judgments of reliability, empirical success, accuracy, and the like can be made, and scientific knowledge may be achieved.

9.7 Objectivity, Rationality, Relativism, and Critique

Should my value-ladenness thesis be rejected because it is incompatible with the rationality, objectivity, progress, and/or empirical success of science, as, for example, Kuhn's early critics have claimed? The value relativity or value ladenness of scientific knowledge is incompatible only with certain questionable philosophical conceptions of the rationality, objectivity, progress, and empirical success of science. The thesis of value ladenness forces us to rethink, not abandon, such norms or concepts. Indeed, the value-laden conception of scientific knowledge opens the way to a fruitful model for a critical theory of scientific argument and the enhancement of rational inquiry. But first, I present some preliminary observations.

Obviously, the thesis of value relativity does not imply an "anything goes" position or one that discounts "the claims" of nature. The work of scientific communities is driven by their attempt to modify their value-laden practices to accommodate natural phenomena that so far resist the forms of description, explanation, prediction, or control they respectively value.

The thesis of value relativity allows plenty of room for objective and rational criticism and decision making. Scientific groups are committed to various epistemic values and aims, as well as the theories, devices, methods, models, techniques, technologies, computations, and other matters that embody their values. Obviously, these commitments constrain inquiry and enable scientific practitioners to question or modify some of their commitments (values) to more effectively realize others. Of course, shared commitments also provide the grounds for practitioners to critically evaluate one another's results and to pursue different lines of inquiry.

Thus, on the view I want to defend, when scientific groups embrace or revise their commitments to epistemic values and standards, typically they have good reasons for those decisions. In some cases, wider practical interests provide good reasons for these epistemic commitments, as in the example of the Bergen school's interest in "the weather," as discussed previously, or they provide contested reasons for epistemic commitments, as in the case of behavioral scientists' standard of workplace stress, as discussed later. On the other hand, epistemic considerations themselves often inform the reasoning or reasons that justify commitments to epistemic standards and values. Some standards are accepted or rejected on the basis of their consistency with, or degree of coherence with, other epistemic standards, theories, or models to which scientific groups are committed. When such inconsistencies arise in scientific practice, there may be no one uniquely rational choice of epistemic standards to resolve the inconsistency, especially in the short run.

Nonetheless, under the pressure of considerations of coherence and consistency, in time the balance of reasons can justify the commitment to a new or revised epistemic standard.

A good example is the debate concerning epistemic standards and the nature of scientific knowledge in the eighteenth and nineteenth centuries, in which rival advocates of inductivist inference, "the method of hypothesis" (or what we call hypothetico-deductive or abductive reasoning), and "the rule of designation" (or what some now call the requirement of novel predictions) defended one or another of these standards as "the" valid basis of all empirical knowledge (Doppelt 1990, pp. 10–18; Laudan 1981, pp. 111–14). The triumph of Newtonian mechanics convinced many natural philosophers that all genuine empirical knowledge depended on a strictly inductive, empiricist standard of inference that excluded, in principle, speculative hypotheses involving unobservable entities. They took it that this standard alone accounted for the great achievements of the Newtonian system and distinguished them from many theories, like the Cartesian vortices, which had been decisively discredited as pseudoknowledge—idle speculations without any sound empirical basis.

Yet the subsequent development of empirical inquiry generated good reasons for rejecting the inductivist methodology as "the" standard of genuine scientific knowledge. By the second half of the eighteenth century, the most successful theories of electricity, magnetism, heat, light, and other phenomena violated the inductivist standard by positing the existence of various unobservable ethereal mediums (or "ethers") to explain these phenomena. George Le Sage developed "the method of hypothesis" to justify the claim that ether theories could be a form of genuine empirical knowledge based on empirical evidence and a reliable mode of inference. On this standard, if a hypothesis or theory entails a large variety of true observational consequences, then it is empirically well founded and counts as genuine knowledge, even if it reaches beyond sense experience to posit unobservable entities and processes (the ethereal mediums). Le Sage cleverly argued that this hypothetico-deductive standard provided a better account of the great Newtonian achievements than the strict inductivist methodology, which was not a standard to which Newtonian knowledge actually conformed. Further, he argued that the method of hypothesis could be rigorously formulated so that it could cohere with the important distinction between spurious hypotheses like those of Cartesian mechanics and others that might constitute genuine knowledge. While conceding that the method of hypothesis was fallible, Le Sage showed that the requirement of infallibility was unrealizable and inconsistent with the recognized achievements of empirical knowledge at that time.

Of course, subsequent developments gave scientific groups good reasons to modify the method of hypothesis by incorporating additional epistemic criteria to further constrain the sorts of hypotheses that might be confirmed as genuine knowledge, such as rule of predesignation (or requirement of novel predictions), the consilience of inductions, considerations of simplicity, and standards of mathematical or explanatory unification. Because epistemic standards and values, theories, techniques of observation, bodies of observed phenomena, and projects of problem solving evolve together, the exposure of inconsistencies and the maintenance of coherence provide scientific groups with powerful epistemic reasons to embrace or revise their commitments to standards of knowledge and epistemic values.

Yet I want to talk of shared commitments to standards or values, because specific groups always need to *decide* how to maintain coherence—what to abandon and what to preserve in the corpus of beliefs and values—if knowledge making is to result. Once such decisions are stabilized in particular epistemic commitments and value-guided practices, the taken-for-granted markers and boundaries of knowledge are maintained, for a time and place, by many social mechanisms that implement normative commitment, without debate and reasoning. This picture supports the practical rationality involved in scientists' commitments to epistemic values and standards, but it is not a picture of disembodied, context-free, formal, or algorithmic rationality.

Scientific rationality, controversy, criticism, and objectivity involve evaluation in which practitioners appeal to some epistemic values or the results attained with them in order to criticize or modify other results or epistemic values.

The thesis of value relativity does not imply the Kuhnian picture of scientific revolution in which one whole set of epistemic values is replaced wholesale by another. The thesis is compatible with continuities, as well as breaks, in the epistemic value commitments of scientific groups, within historical contexts and across them. Thus, even though epistemic value commitments are revised in the development of science, it is possible to identify some powerful elements of continuity and overlap. As a result, current or later theories, methods, values, and the like may represent progress over earlier or outdated ones in the sense that they realize some epistemic values or goals prominent in the past tradition of science, as well as its current practice, that were unrealized in past science; this is a value-relative "progress" because past scientific groups may have embraced and more or less realized other epistemic values that have simply been abandoned by later scientific practitioners. On the other hand, provided that there are good reasons for the changes in epistemic values that have occurred and empirical success in realizing them, the "value-ladenness" of science does not justify relativism or any other view that undermines the possibility of growing scientific knowledge and progress in understanding the natural and human worlds. The fact that scientific inquiry inevitably responds to the practically and normatively salient aspects of nature, theory construction, and reasoning does not undermine the existence of scientific knowledge, reality, and cognitive progress.

The thesis of value relativity abolishes unilinear, cumulative, one-dimensional, and algorithmic conceptions of scientific objectivity, rationality, and progress. It generates the basis for an alternative conception of these norms. I want to conclude by indicating how this thesis opens the way to a critical theory of scientific argumentation that can accommodate "the politics of scientific knowledge," especially in the human sciences, where rival values and interests compete for scientific legitimacy and "objective" grounding in "facts."

On the thesis I've developed, some fundamental arguments over scientific knowledge, among experts and laypersons, may be normative conflicts concerning epistemic values. On the surface, they often appear as arguments concerning the facts: what can or cannot be known through this or that method, approach, or theory; who is or is not in a position to know or speak authoritatively about some subject matter. On my analysis, conflict over the facts of the matter, or scientific knowledge, may embody rival epistemic value commitments and rival practical values. Extreme relativism or irrationalism threaten only if we assume that epistemic value commitments generally are entirely beyond the scope of reason. What forms of reasoning can be exploited by a critical theory of scientific argument grounded in the thesis of the value ladenness of knowledge? Can such a theory provide the means to a more rational conduct of inquiry?

First, such a theory may be used to expose the *rival* epistemic value commitments at stake in scientific controversies that seem to participants, on the surface, to be exclusively about matters of fact or matters of objective scientific knowledge. Such reasoning can reveal that the roots of controversy raise questions concerning rival epistemic standards unacknowledged by the parties to such controversies.

Second, a critical theory may seek to *clarify* the larger political, economic, or cultural interests or values at stake in groups' rival epistemic value commitments. Such reasoning may yield the insight that these provide good practical reasons or bad practical reasons for embracing particular epistemic values, which define the boundaries of science and knowledge in a particular way. The link between a well-established social value and the justification of particular epistemic values may be far less rigid and unilateral than our motivations allow.

Third, a critical theory of scientific argument confronts the root question of whose political or social interests and values are getting embodied in a scientific practice and how to determine whether they provide good or bad reasons for the epistemic value commitments embedded in that practice. Once again, such reasoning may reveal that the epistemic value commitments are rooted in narrow group interests, at the expense of wider, more credible human values or political ideals. Such reasoning could certainly provide good reasons for revising the epistemic value commitments embodied in that scientific practice and for generating new boundaries of knowledge in service to the human values or political values at stake in a given domain of scientific inquiry. Let me conclude with an example of the epistemological politics of value ladenness.

When air traffic controllers went out on strike during the Reagan administration, President Reagan fired them and hired nonunion replacements to do the work and restore the flow of U.S. air traffic (Tesh 1988). The controllers' demands leading to the strike were based on the claim that the controllers were victimized by intolerable conditions of work that generated unusually oppressive patterns of stress in their lives. Because President Reagan's massive action against the controllers' strike was so controversial, Congress held formal hearings to investigate the claims and counterclaims in the case. The controllers had various psychological, subjective, and moral complaints about their work: forced overtime, conflicted demands, disrespect, speedup, lack of control, demands for absolute accuracy, racial and sexual harassment, and others. The advocates for the controllers (the union, Occupational Safety and Health Administration officials) decided that their best strategy in the hearings was to represent the litany of complaints in the scientific, medicalized, objective discourse of "stress." Medical researchers were linking stress to a heightened likelihood of illnesses such as coronary heart disease, stroke, peptic ulcers, and diabetes. By invoking the notion of stress, the controllers' professional advocates hoped to give their complaints and problems a scientific legitimacy and medical urgency to justify their grievances to congressional investigators and to challenge Reagan's repressive action.

Unfortunately, the strategy backfired. Congressional investigators solicited the testimony of scientific experts on stress who discredited the controllers' complaints and supported the Federal Aviation Administration (FAA) position that the complaints reflected individuals' subjective responses to the challenges of work and not a pattern of stressful conditions of work. The experts were behavioral scientists who embraced a paradigm of stress that identifies it by physiological, biochemical, and psychological measurements. The scientists recorded heart rates, measured three stress-related hormones, and utilized psychological tests to assess anxiety levels. On the basis of these standards for determining stress, the experts reached the conclusion that it was empirically unsound to describe controllers' work as an unusually stressful occupation. The objective, scientific facts of the matter spoke against the claims of the controllers and their advocates, and they were discredited by the very medicalized concept of stress that they had invoked. The FAA and the behavioral scientists had measurements, scientific expertise, objectivity, and facts on their side.

Nevertheless, in that context, there was a very different account of stress that *might* have emerged and better served the interests of the controllers, the value of occupational health and safety, and perhaps public safety. For the controllers, stress was something quite different than the body's biochemical reactions to the debilitating emotions and experiences of this work. For them, stress referred to an excessive pace of work and demands for accuracy that left them no time for thought, double-checking, and error self-correction—with thousands of lives at stake in safe or hazardous landings. Stress referred to the experience of staring at a radar screen for long periods of time without breaks or relief, making life-and-death decisions, under intense demands and unrelenting time pressure.

As such, stress was a commonly experienced dimension of the controllers' lives in the work environment, and many noticed that it was followed by common experiences of sleeplessness, irritability, inability to maintain relations of family and friendship outside work, and lower capacity for pleasure and enjoyment. The controllers' common experiences point to different standards for defining the phenomenon of stress and explaining its causes and consequences than the standards used by the FAA's scientific experts.

Of course, these experiences by themselves do not amount to any scientific knowledge of stress, its causes, or its consequences. In this struggle between labor and management, the laborers had an interest— either reducing their stress or getting better compensation for it—that is not translated into the rival epistemic standards and practice of inquiry that might have produced scientific knowledge that would vindicate their claims and their interests. But though the controllers and their advocates were unaware of this avenue of scientific inquiry, we cannot foreclose the possibility of such a successful counterknowledge of stress, based on a disciplined and objective investigation seeking causal links between certain conditions of work, patterns of experience, and negative outcomes beyond the world of work. Informed by a set of epistemic standards and values at odds with those of the behavioral scientists (for example, over the definition of *stress*), such a scientific inquiry might become successful and provide a well-grounded challenge to the facts of the matter concerning stress in such a case.

As things worked out, in the congressional debates between labor (the controllers) and management (FAA), behavioral scientists intervened, in effect if not in intent, on the side of management. Their biological paradigm of stress as a measurable bodily state embodied their specific standards of objectivity, measurement, accuracy, and evidence. The facts that these scientific experts produced discredited the folk-psychological ordinary language reports of the air traffic controllers, to their great disadvantage. For the participants in this conflict of interests or values, scientific authority trumped lay opinion, and fact defeated mere opinion.

This appeared in the congressional hearings as a factual dispute between two parties about one and the same phenomenon: stress in the work of the controllers. The FAA, backed by the expert testimony of the behavioral scientists, achieved a victory—because stress was thought by both sides to be a scientific concept, and the behavioral scientists owned that concept in this context. Thus "the facts" spoke in favor of the interests of the FAA, against the interests of the controllers, and against whatever wider moral values were at stake when Reagan fired all of them and broke the union.

The thesis of the value ladenness of scientific knowledge may provide the basis for a critical theory of scientific arguments for use in cases like this one. This theory awakens the actors to the possibility of the play of rival practical interests and epistemic values in such disagreements, which on the surface appear to be exclusively about "the" facts. The disagreement between the FAA and the controllers was at rock bottom normative, hinging on different interests, rival standards of stress, and potentially antagonistic knowledges of its nature, causes, results, and cure. The controllers paid a terrible price because the epistemological politics at issue were invisible and their interests in and experiences of stress were not embodied in any authoritative scientific voices or knowledge claims. Had the debate been refocused and redirected, in this light, it might have been more rational, objective, and truth conducive. When rival practical and epistemic values are at stake in disputes among scientific actors, or between them and lay actors, the rationality of science is best served if these normative differences are made visible and it is understood that as knowers, we are always already involved in indissolubly linked commitments concerning what to value, how to know, and what to believe. When there are good reasons for these commitments and they inspire an empirically successful practice of science, then practical rationality, the advance of scientific knowledge, and new aspects of nature are linked together in a human narrative of cognitive progress. New facts and truths concerning nature are disclosed, inspired by the expansion of our rational commitments to encompass new practical and epistemic values.

The thesis of the value ladenness of scientific knowledge, as I construe it, opens the way to a richer terrain of normative reasoning concerning knowledge-producing practices. It inspires a critical theory of scientific argument that seeks to clarify the epistemic values and larger social interests or political values that are sometimes at stake in knowledge-making practices and the controversies over them. Because commitments to epistemic values and larger social interests or values are amenable to reasoning and the logic of justification, the thesis of the value relativity of scientific knowledge promises a more, not less, rational practice of inquiry. Beyond that, it can inspire philosophers of science to rethink standard conceptions of scientific rationality, objectivity, realism, knowledge, and progress in ways that accommodate the normativity at play in our understanding of the natural and human world.

ACKNOWLEDGMENT

I thank John Dupré, Harold Kincaid, and Alison Wylie for their helpful comments and generous encouragement. I also thank Paul Churchland for his valuable remarks, and the many insights he has given me concerning philosophy of science, over our years of friendship.

REFERENCES

Doppelt, G. 1978. "Kuhn's Epistemological Relativism: An Interpretation and Defense." *Inquiry*, 21, pp. 33–86; reprinted 1982 in J. W. Meiland and M. Krausz, eds., *Relativism: Cognitive and Moral*, pp. 113–46. Notre Dame, IN: University of Notre Dame Press.

Doppelt, G. 1980. "A Reply to Siegel on Kuhnian Relativism." *Inquiry*, 23, pp. 117–23.

Doppelt, G. 1981. "Laudan's Pragmatic Alternative to Positivism and Historicism." *Inquiry*, 24, pp. 253–71.

Doppelt, G. 1983. "Relativism and Recent Pragmatic Conceptions of Scientific Rationality," in N. Rescher, ed., *Scientific Explanation and Understanding: Essays on Reasoning and Rationality in Science*, pp. 106–42. London: Center for Philosophy of Science Publications in Philosophy of Science, University of Pittsburgh, and University Press of America.

Doppelt, G. 1986. "Relativism and the Reticulation Model of Scientific Rationality." *Synthese*, 69, pp. 225–52.

Doppelt, G. 1988. "The Philosophical Requirements for an Adequate Conception of Scientific Rationality." *Philosophy of Science*, 55, 104–33.

Doppelt, G. 1990. "The Naturalist Conception of Methodological Standards." *Philosophy of Science*, 57, pp. 1–19.

Doppelt, G. 2001. "Incommensurability and the Normative Foundations of a Scientific Knowledge." In H. Sankey and P. Hoynige-Huene, eds. *Incommensurability and Related Matters*, pp. 159–79. Dordrecht: Kluwer.

Friedman, R. M. 1989. *Appropriating the Weather: Vilhelm Bjerkness and the Construction of a Modern Meteorology*. Ithaca, NY: Cornell University Press.

Goldman, A. 1986. *Epistemology and Cognition*. Cambridge, MA: Harvard University Press.

Kuhn, T. 1970a. "Logic of Discovery or Psychology of Research?" in I. Lakatos and A. Musgrave, eds., *Criticism and the Growth of Knowledge*. pp. 1–25. Cambridge: Cambridge University Press.

Kuhn, T. 1970b. "Reflections on my Critics in I. Lakatos and A. Musgrave, eds., *Criticism and the Growth of Knowledge*, pp. 231–79. Cambridge: Cambridge University Press.

Kuhn, T. 1970c. *The Structure of Scientific Revolution*, 2nd ed. Chicago: University of Chicago Press.

Lakatos, I. 1970. "Falsification and the Methodology of Scientific Research Programmes," in I. Lakatos and A. Musgrove, eds., *Criticism and the Growth of Knowledge*, pp. 91–197. Cambridge: Cambridge University Press.

Laudan, L. 1976. "Two Dogmas of Methodology." *Philosophy of Science*, 43, pp. 585–97.

Laudan, L. 1977. *Progress and Its Problems*. Berkeley and Los Angeles: University of California Press.

Laudan, L. 1981. *Science and Hypothesis*. Dordrecht: Reidel.

Laudan, L. 1984. *Science and Values*. Berkeley and Los Angeles: University of California Press.

Laudan, L. 1987. "Progress and Rationality? The Prospects for Normative Naturalism." *American Philosophical Quarterly*, 24, pp. 19–31.

Laudan, L. 1988. "The Rational Weight of the Scientific Past: Forging Fundamental Change in a Conservative Discipline," unpublished manuscript.

Longino, H. 1990. *Science as Social Knowledge*. Princeton, NJ: Princeton University Press.

Meiland, J. 1974. "Kuhn, Scheffler, and Objectivitiy in Science," in *Philosophy of Science*, 41, pp. 179–87.

Meynell, H. 1975. "Science, the Truth, and Thomas Kuhn." *Mind*, 84, pp. 79–93.

Morrison, M. 2000. *Unifying Scientific Theories. Physical Concepts and Mathematical Structures*. Cambridge: Cambridge University Press.

Partington, J., and McKie, D. 1937. "Historical Studies on the Phlogiston Theory, I. The Levity of Phlogiston," *Annuals of Science*, 2, pp. 361–404.

Sankey, H. 2000. "Methodological Pluralism, Normative Naturalism and the Realist Aim of Science," in R. Nola and H. Sankey, eds., *After Popper, Kuhn and Feyerabend: Recent Issues in Theories of Scientific Method*, pp. 211–29. Dordrecht: Kluwer.

Scheffler, I. 1967. *Science and Subjectivity*. Indianapolis, IN: Bobbs-Merrill.

Scheffler, I. 1972. "Vision and Revolution: A Postscript on Kuhn. *Philosophy of Science*, 39, pp. 366–74.

Shapere, D. 1964. "The Structure of Scientific Revolutions." *Philosophical Review*, 73, pp. 383–94.

Shapere, D. 1966. "Meaning and Scientific Change," in R. Colodny, ed., *Mind and Cosmos: Essays in Contemporary Science and Philosophy*, pp. 41–85. Pittsburgh: University of Pittsburgh Press.

Shapere, D. 1971. "The Paradigm Concept." *Science*, 172, pp. 706–9.

Shapere, D. 1982. "The Concept of Observation in Science and Philosophy." *Philosophy of Science*, 49, pp. 485–525.

Shapere, D. 1984. *Reason and the Search for Knowledge*, Boston Studies in the Philosophy of Science 78. Dordrecht: Reidel.

Siegel, H. 1980. "Epistemological Relativism in Its Latest Form." *Inquiry*, 23, pp. 107–17.

Tesh, S. 1988. *Hidden Arguments: Political Ideology and Disease Prevention Strategy*. Trenton, NJ: Rutgers University Press.

Zammito, J. 2004. *A Nice Derangement of Epistemes*. Chicago: University of Chicago Press.

TEN

CONTEXTUALIST MORALS
AND SCIENCE

Harold Kincaid

AS THE OTHER CHAPTERS IN THIS VOLUME ATTEST, VALUES ARE SURELY
involved in science. Yet the interesting and still controversial question
remains: What moral to draw from this fact? In turn, what morals are
drawn depend on why and how values are implicated in science. Here
controversy remains as well.

In what follows, I hope to make progress on these problems by first
carefully distinguishing various roles values might play and the kinds of
arguments advanced, both for the value ladenness and value freedom of
science. I then argue that arguments on both sides of the issues rest on
untenable assumptions about traditional epistemology as a coherent enterprise. Dropping those assumptions opens a space for showing that science can be value laden and yet robustly objective.

The chapter is organized as follows. Section 10.1 sorts out issues,
sketches some standard arguments both for and against value ladenness,
and then describes epistemological assumptions common to both and
argues that those assumptions are misguided. Section 10.2 then presents
an alternative picture of values and science without those assumptions.
Section 10.3 applies that picture to debates over values and objectivity in
economics and medicine. The issues here are large and my discussion
brief. What follows aims to sketch a relatively novel view and show it
worth pursuing, not provide a detailed defense.

10.1 Issues, Arguments, and Misguided Presuppositions

The introduction to this volume outlines four key questions about values in science: What kinds of values are involved—moral, political, epistemic? How are they involved—inevitably or contingently? In what part of science—questions asked, confirmation, explanation—are they involved? And what does this tell us about science?

A traditional view is that values are only contingently involved, and that when they are, they make science subjective, because value claims are subjective in a way that factual claims are not. The standard picture is opposed by the current majority opinion in the science studies literature, which concludes that values are essentially involved. Opinion then is divided on what this implies for the objectivity of science. Some argue that science should be partisan. Others (Longino 1990, 2002) defend objectivity but think our notion of objectivity has to be revised to allow moral and political criticism.

Both the traditional value-free science position and the social constructivist view that values are inherent are wrongheaded, I shall argue. Both rest on a mistaken presupposition about the nature of knowledge and justification that underlies key arguments on each side. I want to first sketch the mistaken ideas about knowledge and then show how arguments on both sides presuppose those mistaken conceptions.

The approach to knowledge and justification that I favor (and assume, not defend) goes by the name of "contextualism."[1] In its negative form, contextualism makes claims of the following sort:

1. We are never in the situation of evaluating all of our knowledge at once.
2. Our "knowledge of the world" is not a coherent kind that is susceptible to uniform theoretical analysis.
3. There are no global criteria for deciding which beliefs or principles of inference have epistemic priority.
4. Justification is always relative to a specific context, which is specified by the questions to be answered, the relevant error probabilities to be avoided, the background knowledge that is taken as given, and so on.

Thus contextualism rejects various foundationlist approaches in epistemology, both in the sense that it denies that there are special classes of belief that undergird all others and in the sense of denying that epistemology provides a priori conceptual standards for knowledge and justification.[2] Put in a positive vein, contextualism claims that all evaluation is local and empirical. Standards for justification build on what is already known in a given situation and carry no more and no less weight than the empirical information on which they are based.

Related contextualist ideas include:

> *The local nature of scientific realism*: Contextualism has negative and positive morals in debates over scientific realism. The negative moral is that arguments for and against scientific realism cannot proceed by evaluating all of science at once, by appealing purely to formal or methodological grounds, and without invoking substantive empirical background information. So *global* realist and antirealist arguments are equally misguided. Put positively, arguments over realism must proceed by assessing specific theories or fragments thereof, given what else is known in the relevant context. Such arguments are as much scientific as philosophical. Their conclusions may show us that we both know much in some cases and very little in others.
>
> *Postnomonological-deductive work on explanation*: The nomonological-deductive (N-D) model of explanation embodies the search for purely formal criteria for evaluation. An overwhelming series of objections have shown that the N-D model can be no such thing. Work since then has pointed to the pragmatic factors involved in explanation that vary according to context; for example, audience and assumed background knowledge are essential. Contextualism naturally endorses these claims about explanation (Kincaid 2004).
>
> *Antiessentialism about theories*: Theories are not unified, monolithic entities, and they are not the whole of science. From the work of Kuhn (1962), work in the history and sociology of science (Beller 1999), and work by Cartwright (1999) and others on the role of models in science and so on, we have good reason to believe that "theories" have diverse interpretations across individuals and applications, are often not a single axiomatizable set of statements, and involve differing kinds of extratheoretical assumptions and devices in the process of explaining. From much of that same work, we also learn that skills, material culture, and the social organization of expertise are essential to science.
>
> *The restricted role of conceptual analysis: finding the necessary and sufficient conditions for concepts is often misguided and uninformative.* There is good reason to think that concepts are generally not represented by necessary and sufficient conditions. Even if they were, it is often unclear what light an analysis based on a group of philosophers' linguistic intuitions will shed on science or reality. Conceptual

clarification, of course, has a place but only as part of empirical investigation, not as substitute for it.[3]

Other approaches that have much in common with contextualism include Fine's (1986) "natural ontological attitude," which tries to reject both sides of standard debates on scientific realism by arguing that they bring unwarranted philosophical interpretations to the science; naturalized epistemology of the sort advocated by Quine, which denies that there is any purely epistemic project separate from science (much naturalized epistemology makes the weaker claim that science is relevant); Maddy's (1997) naturalism about mathematics, which tries to show that the main questions about the rationality of controversial set theoretic axioms are not decided by taking a prior position on the existence of a separate mathematical reality; Sklar's (2000) look at ontological elimination moves in science, which concludes that the best arguments are local and contextual and do not appeal to global philosophical principles, a view explicitly endorsed by Einstein vis-à-vis his use of verificationism; various kinds of minimalism or deflationism, for example, arguing that we need no theory of truth over and above the trivial "P" is true just in case P; recent work on cashing out objectivity in science (Douglas 2004), which finds that that there are numerous different scientific virtues being appealed to that vary according to the scientific situation and that have rather different implications about values and social processes in science; and the pragmatist tradition past and present (Putnam 2002), which has probably influenced all of the preceding. These contextualist ideas will become more concrete as we trace their implications for values in science in the rest of this chapter.

With this brief sketch of contextualism, I now turn to analyze arguments both for and against the inherent value ladenness of science. The upshot shall be that they presuppose precisely the picture of knowledge that contextualism rejects.

A very influential set of arguments sees value ladenness in science arising from the underdetermination of theory by data (Longino 1990). The idea behind underdetermination is that even if we had all possible data, there would still be multiple hypotheses compatible with the data. In the introduction to this book, this claim was illustrated by means of the curve-fitting problem: from a finite set of data, it is possible to extend the curve in indefinitely many ways consistent with a specific,

given set of data. In short, finite data are always too impoverished to decide between rival generalizations that go beyond the data. Another
form of the underdetermination arguments proceeds from the claim
that there is no logic of science—no algorithm or criteria that uniquely
tell you what to believe, given the data. In terms of the curve-fitting example, the argument would be that there is no guaranteed way to pick
one right curve from the many curves consistent with the data. The upshot in either case is that something other than data and rules of inference must be involved, namely, values, both epistemic and nonepistemic. Values, the conclusion is, are involved in the very heart of good
science in the confirmation process. Longino, for example, provides numerous case studies from the natural sciences where data and method
alone do not dictate conclusions.

From the contextualist perspective, these arguments fail because of
the way they set up the problem. Comparing "all the data" with "all the
possible hypotheses" is an incoherent situation we are never in. "All the
data" is not something we can ever have, nor do we ever approach data
and hypotheses innocently without background knowledge—about the
importance and relevance of some data over others, about the possibly
relevant hypotheses, and so forth.

Only if we ignore such background information does the underdetermination claim look plausible as a claim about what must happen.
The contextualist point is that we have no good reason to do so—no
good reason to assess data and hypothesis somehow denuded of our relevant background knowledge. Of course, underdetermination is possible but not as a global problem, however, only as a *local* one. We sometimes believe this particular data set at this time or cannot decide
between these hypotheses, given what we now know. However, that
kind of underdetermination is a call for more investigation, not for invoking moral or political values. Thus underdetetermination is not necessary, and thus neither is the role of values in confirmation.[4]

Related arguments come from the defenders of social constructionist views of science. They sometimes argue that the historical evidence
shows there is no logic of science and thus it is inevitable that social
factors—and thus values—are involved in confirmation. So, McCloskey
(1994), for example, points to failed searches for a logic of economic ap-
praisal and forthwith concludes that all economic argument is rhetoric,
thus opening a wide space for values in the heart of science.

The reply to this reasoning is the same as before: Such arguments assume that knowledge comes from a special set of rules with an inherent epistemic status as knowledge conferring. The contextualist denies there are such rules and denies that knowledge requires them. We have good reason to think we know some things. Those reasons are stronger than any we might give for claiming that knowledge requires a logic of science. Although it is conceivable that values might fill the gap between inference rules and our conclusions, there is no reason they must; we often have a wealth of nonnormative background knowledge to do so. In our curve-fitting case, for example, we may have good prior reasons to think the relation is linear rather than quadratic and thus good empirical reason to throw out many possible lines through the data. Science driven by a logic of science and science driven by values and social factors are not the only two possibilities.

The standard arguments for the inherent value freedom of science can be similarly disposed of. Most such arguments take the form of moral skepticism and try to show that moral claims must have a fundamentally different status than factual scientific claims. So, since Hume, a certain kind of moral skepticism has been based on the claim that moral truths cannot be derived from purely factual truths. Similar sentiments are expressed in the claim that moral truths do not supervene on nonmoral or natural truths and thus are suspect, or that moral facts are not natural facts and thus "queer" (Mackie 1977).

The literature on these topics is long and complex. Although I do not pretend to address all that work and the issues in any detail, I can sketch a unified contextualist response. All these arguments rest on the assumptions that there are special kinds of inference relations or beliefs that ground our knowledge of the world and that showing an absence of those allows us to dismiss entire domains of putative knowledge. These assumptions the contextualist denies. However, they turn up again and again in the arguments deployed by moral skeptics. Let me survey some of the better known variants.

Take the Humean argument first. Maybe moral claims cannot be derived from nonmoral ones.[5] But then biological claims cannot be derived from just those of physics; this holds for every special science vis-à-vis those sciences more basic. Do we then want to put these areas in limbo as well? Many have been forced to some such position by this logic. Yet for the contextualist, these radical conclusions are reason to question the criterion invoked. Why assume that for some domain of belief A, those beliefs are legitimate only if they follow by logic alone from some more fundamental domain? We should ask first why the domain in question is really so fundamental; why isn't this just begging the question against the defender of the objectivity of morals? Moreover, even if we grant some set of fundamental facts, why is logical entailment the essential relation? Our biological knowledge has many inferential and evidential ties to our physical knowledge, but they are not logical inferences from pure physics. Why hold morality to a higher standard?

A similar diagnosis confronts the arguments that moral facts do not supervene on natural facts. Again, note that this type of argument threatens to generalize to areas other than morality. Psychological facts arguably are not fixed by the neurobiological (because meaning is in part social), and they are certainly odd from a physical point of view. Something as basic as the chemical facts about bonding are not at this point fixed by the facts of quantum mechanics (Hendry 2003). That is again reason to give us pause. Furthermore, what we take to be the "facts," "kinds of facts," and "natural facts" are not simply intuited but intimately connected to what we think we know. We should thus expect the various kinds of facts to have complex interconnections to each other, just as do the various special sciences.

Much of the debate over the status of morality is put in terms of whether there are moral facts. From the contextualist perspective, this is the wrong way to look at the issue. Put this way, it tries to evaluate entire domains of discourse in a single swoop on conceptual grounds. That cannot be done. Judgments about what kinds of facts exist are parasitic on confirming specific claims in those domains, not vice versa. Another way to put the point is in terms of truth. The contextualist denies that knowledge and justification have an inherent nature. It is a short step to a similar claim about truth, namely, that there is no prospect for a substantive theory of truth. Deflationism (Horwich 1990), as this view is called, holds that "x is true" if and only if X and denies that there is anything more in the way of a theory of truth possible or needed. That removes the demand to find some special kind of fact that makes moral claims true. The queerness of moral facts does not arise, and the debates have to be moved to the plausibility of specific moral claims. No uniform treatment should be expected.[6]

The search for a special set of facts gets its bite from the implicit assumption that some beliefs must have an inherent privileged epistemic status if there is to be objectivity in morality—that there must be a unique set of knowledge-conferring rules if morality is to have objectivity. So Korsgaard (1996) claims that the essential question is "what grounds morality?" and then goes on to argue that facts and reflection cannot do the job but that autonomy can. Mackie (1977), in arguing for the "queerness" of moral facts, explicitly claims that if there is objectivity in morals, there must be a special moral sense. Bernard Williams (1985) argues that we should take science realistically because its success is explained by its truth, unlike morality. Harman (1977) has a similar argument, namely, that we do not need moral facts to explain moral beliefs, but we do need scientific ones to explain scientific belief. Joyce (2001) argues that morality, like the phlogiston theory of combustion, rests on a false presupposition and thus the whole enterprise falls.

In each case, there is a key implicit assumption that the contextualist rejects. In Korsgaard's case, the assumption is upfront: Morality has to be grounded on some special reason-giving foundation. The foundation she chooses—Kantian autonomy—has long been thought to exhibit exactly the problems contextualists would expect. A standard criticism of Kant's use of the categorical imperative is that the rule is not really purely formal or, in other words, universalizability doesn't get us moral conclusions without making substantive moral assumptions. This is just what contextualism thinks dogs most attempts to decide what to believe on the basis of formal rules—vacuousness or disguised use of nonformal considerations. Bayes's theorem as a logic of confirmation arguably has these traits, for example (Day and Kincaid 1994). Korsgaard has surprisingly little to say in response to these standard criticisms.

Williams's and Harman's arguments both invoke a general rule of inference—inference to the best explanation—that they believe makes us realists about science and antirealists about morality. They put their faith in a global argument for scientific realism—using the allegedly foundational inference rule "inference to the best explanation"—that the contextualist finds flawed for obvious reasons. The problem is this (Day and Kincaid 1994). Inference to the best explanation requires some account of explanation. Purely formal accounts of explanation have well known problems. Nonformal accounts, such as explanation as the citing of causes, rest on substantive and domain-specific knowledge. So inferences to the best explanation are no better than the substantive assumptions they make and have to be evaluated in those terms. Defending science tout court on the grounds that its truth best explains its success is much too thin a story. We know that successful theories have turned out to be false and that some respectable science in the past is best seen as a social construct with no claim on truth. We know that social processes are involved in all science, so the explanation of scientific success can never just be its truth. Until alternative explanations of success are ruled out, no conclusion can be drawn about the best explanation of science. But making these more specific arguments will be a case-by-case affair.

Thus Williams ascribes to science something it can never do and, in the process, evaluates all of science in one fell swoop. No wonder morality looks weak by comparison. Yet if we look at particular pieces of science and particular pieces of moral reasoning, we can find that both involve social processes and claims that are well confirmed. The stark contrast disappears. This is, of course, compatible with finding that science is generally in much better shape than morality, but this has to be a generalization from specific cases, not an a priori inference from "the nature of science" and "the nature of morality."

Mackie's argument rests on the assumption that if there is moral knowledge, there must be a moral sense. This assumption is, in turn, an instance of the foundationalist view that there are certain sets of beliefs that are by their very nature more certain than others, that is, those that are sensed rather than inferred. Obviously, this is an assumption that contextualism denies.

Joyce's error theory rests on the claim that if a theory rests on a false presupposition, that suffices to show that no knowledge is produced by the theory. This is another misguided attempt at a purely formal test to evaluate entire domains. There are several ways to see this. Consider first the phlogiston example. Kitcher's (1978) careful work on precisely this example illustrates the actual story of how phlogiston theory developed into the discovery of oxygen; he shows that the reference of phlogiston changed as the experimental context developed, producing knowledge at some points and not at others. To put the criticism more generally, any look at real scientific practice will discover numerous false assumptions at any given time, a conclusion Harman (1986) defends for real-life reasoning in general. If an error theory of morality is plausible, based on Joyce's reasoning, than an error theory for every scientific domain is plausible. Of course, it is not.

Lurking behind all of these arguments for moral skepticism are some specific assumptions about objectivity, and framing the issue in this way is another route to seeing the contextualist point. The assumption is that for a discourse to be objective, there must be a set of formal criteria it must meet. Or to put it slightly differently, "objectivity" has a formal definition. The contextualist claim is that we do not need and cannot get any such thing. We have various methods for identifying what is objective and various different things we are trying to achieve in doing so. This latter claim has been nicely illustrated by Douglas (2004).

So these arguments for the subjectivity of values and thus for value-free science presuppose the picture of knowledge and justification that contextualism rejects. Rejecting that picture leaves space for moral knowledge and thus for the relevance of values to science in a way that does not threaten the latter's objectivity.

10.2 The Place of Values in Science

What, then, is the place of values in science on the contextualist view? There are three *compatible* answers to this question: (1) Their place is the same as any other kind of claim, (2) the question is misguided, and (3) they largely have no place in science. Let me explain.

The place of values in science is the same as any other type of claim on the contextualist view, because no type of claim has an inherent or automatic epistemic status or, to borrow Williams's phrase, "value claims" are not an epistemic natural kind—any more than what we call "our factual beliefs" have an automatic status as probable, improbable, and so on. Just as is the case for "all knowledge," there is no principle of adequacy that informs all moral knowledge, and we are never in the situation of evaluating all our moral beliefs at once.

The upshot is that objective-subjective and rational-irrational types of distinction do not correspond to the factual beliefs–value beliefs distinction. Some value claims may be far more compelling than scientific claims in that they have far better reasons on their behalf. Nothing about the nature of knowledge or of the moral (because on the contextualist view, they have no nature in a sense) precludes this.

From this conclusion another follows: Moral beliefs might be relevant evidence in assessing scientific belief. Some moral beliefs are well supported. If we accept some kind of principle of total evidence, then we cannot rule out a priori that moral judgments are relevant to the confirmation of scientific claims.

One powerful piece of evidence that moral claims do have such a role comes from the collaborative nature of scientific research. Much scientific research is sufficiently specialized and distributed that no one individual can understand, not to mention vouch for, the final product as a whole. In such cases, trust is crucial, and the credibility of results depends on the character of the investigator who produced them. However, making reasonable judgments about who is trustworthy is making moral judgments.[7]

This fact gives us a kind of "indispensability argument" of the sort that Quine proposed for evaluating mathematics. If there is some domain where we believe we have knowledge, and beliefs of kind M are essential for arriving at that knowledge, then we have some reason to think that beliefs of kind M are knowledge.[8] Once we grant that science is a thoroughly social process and that justification is collective, these considerations are hard to gainsay.

This puts moral considerations at the heart of science in the confirmation processes. There are, of course, other areas where values also play a legitimate role, but these are somewhat less controversial. The contextualist picture of explanation identifies numerous contextual factors that must be set in the process of explaining. Explanatory questions have an implicit contrast class (e.g., why did Adam, rather than Eve, eat the apple?) and standards for relevant answers (e.g., psychological versus theological). Sometimes these contrasts are dictated by pure scientific considerations, but not always: Our interests and thus values frequently make some questions and answers more relevant. Note, however, that nothing here implies that further values are implicated, once the question and kind of answer are set. As a result, contested moral values—those that may be bad candidates for moral knowledge—can play a legitimate role without compromising the science if their place is explicit.

Now for the other two possible answers to our question about the place of values, and here we qualify the prior claims that this question is misguided and that values have no place. Because contextualism denies an inherent epistemic status for moral claims, then there is no all-purpose role that values play in science—at least none that we can decide a priori. At best, we can generalize from concrete analyses of moral claims in scientific arguments. So the question is in this sense misguided.

Finally, the third reply to our questions about the place of values— namely, that they have no place—is not hard to see. The proper role of values has to be evaluated by experience. Nothing a priori rules out that moral claims are often irrelevant to scientific belief: Just as there is nothing about the essential nature of science and morality that entails moral judgments cannot be positive evidence for scientific conclusions, there is nothing that entails they must be. We certainly can cite a host of cases in the history of science where the moral and political values of scientists were a genuine hindrance.

Actually, we should be suspicious that even these three different conclusions about the place of values are too general. The contextualist denies that the basic categories of traditional epistemological theory pick out useful natural kinds. One way this suspicion surfaces is in doubts about "theories"—doubts about them as monolithic, unified entities that are the core of science. Actual science is more multifaceted, for it involves much more than theories, and its theories are often diverse and independent sets of claims, models, idealizations, and so on. So the place of moral claims need not be uniform across these different facets of science (and there are also lurking implications here for the project of finding a moral theory that is taken to exhaust the practice of morality).

Thus, discussing the place of values in the abstract can go only so far. I thus turn next to look at values in two specific scientific settings— in development economics and in medical classification.

10.3 The Value of Knowledge about Poverty and Health

Development economics seeks to explain why some countries have become prosperous developed economies and others have not or have become so belatedly. To explain these things, there must be some measure of economic development. And to provide that, development economics must draw the line between economic and other activities. However, differences in drawing the productive-nonproductive divide are strongly associated with political ideology and not dictated by economic logic (see Boss 1991). Marxists have argued that various kinds of advertising and defense expenditures bought in an open market are nonetheless nonproductive activities, whereas conservatives have said the same about activities directed to influencing government policy, again when those activities involve services (e.g., lobbying) purchased in a market.

Values are thus indeed involved at this fundamental level in economics, but their role is various with diverse implications. In particular, we should distinguish between their role in (1) *confirming* accounts of the causes of development and underdevelopment, (2) the kinds of *explanations* given, and (3) the use of development economics to make applied *policy recommendations*. These are different facets of science, and there is no reason to think they should be treated the same.

The most natural way to identify economic activity and hence to measure economic development is by reference to markets. Economic activity is that undertaken to buy or sell goods or services on a market. The total economic product of a country is the sum of items sold.

Although this definition is intuitive, it hides many ambiguities. To begin to see the problems, ask if the goods in question must actually be sold on a market or only *could* have been. Under the former criterion, the carpenter who makes two identical chairs and sells one in a store and takes the other one home has contributed only one economic product. That seems arbitrary. The system of national accounts (SNA)—the standard gross domestic product (GDP) accounting system—deals with this problem by including bartered goods and goods produced and consumed directly by households.

Yet the SNA excludes services (e.g., cleaning toilets) produced by households, even when the same service can be purchased in a market. It is not hard to see how values might intrude here in a negative fashion. As feminists have argued, not counting household services mean disproportionately discounting the activity of women. Women are considered to belong in the home by their nature, and it is men who bring home the bacon on their way back from productive work.

Values thus surface in this work in several ways with different implications. There are arguably multiple, equally valid ways to draw the economic-noneconomic distinction. Values may motivate drawing the distinction one way or another; how we draw the distinction also may have value implications. But, following our contextualist morals, we should not jump to quick conclusions about inherent objectivity or subjectivity.

Policy decisions based on the SNA do inherently make value judgments. They undervalue the labor of women and overvalue the labor of men. They also equate well-being with a certain restricted notion of material well-being, another value decision. This results in International Monetary Fund (IMF) policies that do not count the negative effects on households and women. But note that they inherently undervalue women only because of existing social relations, not because of the inherent nature of economic theory.

However, our moral reasons for finding SNA inadequate are not subjective opinions. There are no good reasons to think that the labor of women is less valuable than that of men simply because it is done by women, nor is there any good reason to think that human well-being is just material well-being. People may have thought or still think otherwise, but the arguments against them are overwhelming. We can be about as certain about this as we can be, for example, that HIV causes AIDS (which some experts still deny). This is a case where moral considerations are scientifically relevant and reasonable.

Yet, in another sense, values need not and should not be involved in testing and confirming claims about economic development. Although there are multiple ways to draw the economic-noneconomic line, *once the distinction is made*, then hypotheses about the causes, extent, and so forth of development in this sense can be confirmed by usual methods. If we are explicitly interested in the development of the market economy and consistently study the causes of growth in market production, then the ensuing causal claims are value free in the relevant sense. No moral arguments need to be made one way or another to do so. If they were involved, we would suspect bias.

Of course, there may be some gray areas over what counts as a market. If our way of drawing that line systematically excludes women, that might well be a subjective value-laden assumption. Alternatively, if we do count some nonmarket processes as contributing to the GDP but do so in a way that again systematically excludes women, bias is probably at work. Feminist economists have argued that precisely this happens in development economics.

Let me describe next two other ways that values are involved in development economics that have a more ambiguous epistemic status. The first concerns further issues about measuring growth; the second, issues about the causes of growth.

Any measure of growth will count most market transactions as contributing to the total product ("most" because some might be counted "unproductive"). But market values are determined by the fundamental factors influencing supply and demand, among which are the preexisting distribution of wealth and tastes. When the causes of growth are identified or the effects of government policy assessed, the distribution of wealth and tastes is taken as fixed. So these apparently morally neutral assessments are made on the presumption that the existing distribution of wealth, for example, is going to remain the same.

Is that a moral assumption? Contextualists are suspicious that scientific theories have transparent and constant meaning in all contexts. A similar thesis seems plausible about "moral language" as well, and this seems a case in point. If the IMF or World Bank imposes requirements on developing countries that are the "best for long-term growth" and do so taking the distribution of wealth as given, they may be taking a political stand. Unequal land distribution is a chronic problem in developing countries, and there are good reasons to think that the right kind of land reform might significantly contribute to efficiency and thus growth. Part of Taiwan's economic success, for example, may be in part due to its land reform policies. So making policy recommendations on the assumption that the distribution of wealth remains unchanged is implicitly condoning that situation and, in that sense, taking a moral stand. That moral stand does not entail that values are involved in the *evidence* for the claim that a policy will have certain effects, given the distribution of wealth. So the deep moral disagreements about the proper distribution and redistribution of wealth need not imply bias in the causal claims made. Yet some kind of bias does seem involved, especially if the presuppositions about fixed distributions of wealth are not made explicit to an audience that would find them suspect.

A second and similar place for values surfaces by a similar route in general theories of growth. Explanations of growth and stagnation tell causal stories. Those stories make assumptions in the kinds of questions they ask and in the kind of answers taken to be relevant, as do all explanations. Those assumptions, however, have indirect connections to moral and political controversies but are not explicitly identified as such. Examples of this situation abound in the development economics literature, but let me focus on one particularly informative instance.

The history of development economics is filled with claims of the form "the main cause of growth is X." So, according to Lucas (1998) it is human capital, to Krueger (1984) it is free trade, and to Barro (2001) the rule of law. Let's grant that these are generally positive causes of growth. The difficulty comes in the assertion that they are the "main" causes. There is little sense in this idea. Development is a complex causal process that involves many causally relevant components, many of which can be necessary prerequisites. As such, there is no obvious, unique metric for saying that some factor is the most important simpliciter.[9] What we can do is use our background interests and knowledge to focus on particular aspects of the causal process. The continued light of the sun is a cause of development but one we have no control over and thus ignore. Yet that is not because it is less important; compare the sun–no sun situation to the free trade–autarky contrast, and I bet you will see the former has considerably more influence on growth in an intuitive sense!

Values are lurking here in the choices of what variable to describe as "main" — in deciding what part of the complex process we focus on. Human capital and free trade fit nicely with the political philosophies of many economists that emphasize minimal government and individual responsibility. However, as we saw before, the distribution of wealth is also a part of the complex causal web. There is an implicit value judgment in the decision not to focus on its role.

My conclusion is that values are involved in development economics. But they have no uniform effect. They may indicate bias and bad science; they may be an objective part of good science; they may not be involved.

I want to turn now to argue similar theses about medicine. Earlier, I looked at the very definition of economics, and here I look at an equally fundamental notion in medicine, namely, health (putting aside the more complex issues of mental health).

There is a long-standing debate, inside medicine and out, about how to define disease and about whether such definitions are value free. Defenders of medicine as a science look for value-free notions on the premise that they are essential for science. Radical social constructionists argue that values permeate our notions of health and that medicine is a paradigm illustration that science is politics. Both sides are confused on my view.

The two predominant attempts at value-free notions of health are the biostatistical theory (BST; Boorse 1977) and evolutionary functions approaches. The biostatistical theory holds that disease is deviation from species-typical functioning; disease is deviation from the average. In the evolutionary function view, disease occurs when an organ is not performing the job that allowed it to evolve via natural selection.

Both definitions face notable problems. Having dental caries at the rate typical for humans is perfectly healthy, according to the BST. Our apparent natural taste for fatty foods probably had positive selective value as we evolved as hunter-gatherers on the African savanna and thus cannot be unhealthy on the evolutionary view, even if it causes arteriosclerosis and obesity. This is not surprising from the contextualist perspective, because these accounts are predicated on the search for necessary and sufficient conditions for applications of the concept, a doomed project from the start.

The deep problem here seems to be that notions of human health have an essential tie to our conceptions of human welfare (and other approaches to health build that in directly to the definition). However, it is very hard to argue that conceptions of human well-being do not involve value assumptions—witness the recent controversies over giving growth hormone to short-statured individuals or over whether deaf children should be taught to speak.

Although values are involved in the fundamental concepts of medicine, I think this case also illustrates my contextualist theses that the role of values is various, that value involvement need not imply subjectivity, that it may often be the case that values should not be involved, and that sometimes they should.

The role of values varies in part because a value-laden notion of health does not imply that values are involved everywhere in medicine. Although the conception of disease may be normative, the methods we use to test for the presence of the disease need not be. If I need a normative conception of human well-being to decide if short stature is a medical condition needing treatment, I do not need one to measure height or to tell whether height increases with treatment by growth hormone.

When values are involved, objectivity need not be compromised for two reasons. As we just saw, once a definition of health is accepted, then ordinary, nonnormative methods can be used to determine what effect a treatment has, for example. Moreover, the presence of human values does not mean that some judgments of ill health cannot be well confirmed. Childhood cancer is as clear a case of ill health as shortness of stature is an ambiguous one.

Finally, when values surface in diagnosis and treatment decisions and those values are contentious—when they are closer to "shortness is a medical condition" than to "childhood cancer is an illness"—then values can be bad for medical science and something that ought, both ethically and scientifically, to be avoided. The literature on ambiguously gendered babies provides many such cases.

I want to conclude this discussion of values and notions of health by noting another way values may be involved that closely parallels the role of values in theories of economic growth. We saw that values were implied there in the process of explanation—in setting the contextual parameters needed for defining causes. The same situation occurs in identifying the causes of disease.

The notion of a "genetic disease," so prevalent in the current literature, is a good case in point. The role of genes and environment in disease is much like that of the factors in development; there is no clear sense in talking about one as a "bigger" cause (Kaplan 2000). It proves irresistible to many to do so, but sophisticated biologists grant that the whole notion is misguided. Focusing on genes is certainly a legitimate tactic in understanding the causal processes of disease, yet genes are always only part of the story. In the right social and political context, restricting attention to genes again makes implicit value judgments, for example, that responsibility for disease and the relevant target for its elimination is the individual rather than unhealthy socioeconomic conditions.

10.4 Conclusion

I have made an attempt to reframe the debate over values in science by rejecting some traditional epistemological assumptions behind it. No doubt much work needs to be done to make this alternative approach clear and plausible. Equally important work is called for in applying it to cases; the short discussions in this chapter show that there are many questions to pursue, and raising many questions is a sign of a progressive research program.

ACKNOWLEDGMENT

I would like to thank Ted Benditt, Don Ross, Scott Arnold, Maureen Kelley, and Nathan Nobis for helpful comments on earlier drafts of this chapter.

NOTES

1. There are two uses of contextualism currently in vogue. David Lewis and others have developed a view about the semantics of the term *knowledge* such that it contains essential reference to context. This may entail various of the claims I attribute to contextualism about knowledge, but my version does not accept (1) and (2). Those are parts of the analytic epistemology tradition of finding necessary and sufficient conditions via a priori conceptual analysis that contextualism, as I defend it (and as does Williams), rejects. Timmons's (1999) work on moral contextualism is in the same boat. Some of his claims about the nature of moral justification I would endorse, but the contextualism I am defending would reject much of the philosophical machinery he employs and the framing of debates in metaethics.

2. Contextualism in this sense is a stronger position than some varieties of naturalized epistemology that make empirical information and science relevant but in the end defend criteria of knowledge and justification via conceptual analysis that is a priori (e.g., Goldman).

3. The best argument of this sort is perhaps Stich (1990).

4. Longino, who relies on these kinds of arguments, denies that withholding judgment to avoid values is a real option and thus seemingly endorses the "must" interpretation, though she grants that something other than normative values might close the gap between data and hypotheses (2002, p. 50).

5. An easy argument for the opposite conclusion I learned from Frank Thompson but have never seen in print: A moral conclusion is one where a moral attribution is claimed to be true or false. Let that be Q. Then from any nonmoral claim P, it follows that P or Q. Of course, various gyrations are possible in response, but the fact that they are necessary should be enough to jar your intuitions.

6. Arguments with a similar force are offered by Putnam (2002) in claiming that the fact-value distinction need not imply a metaphysical distinction.

7. Of course, critics can try to couch out these judgments as nonvalue predictive judgments about personality and behavior. The contextualist will resist that reductionist project.

8. This is a much weaker form of the argument than Quine gave, which asserted what we must conclude and did so about ontological interpretation. See Maddy (1997) for a careful discussion along these lines.

9. A standard rhetorical strategy in the social sciences literature is to use explained variance as the measure of causal importance. This is another instance of the longing for purely formal criteria that is misguided, because R2 can be a terrible measure of causal importance.

REFERENCES

Barro, R. 2001. *Determinants of Growth*. Cambridge, MA: MIT Press.

Beller, M. 1999. *Quantum Dialogue: The Making of a Revolution*. Chicago: University of Chicago Press.

Blackburn, S. 1993. *Essays In Quasi-Realism*. Oxford: Oxford University Press.

Boorse, C. 1977. "Health as a Theoretical Concept." *Philosophy of Science*, 44, pp. 542–73.

Boss, H. 1991. *Theories of Surplus and Transfer: Parasites and Producers in Economic Thought*. London: Unwin Hyman.

Cartwright, N. 1999. *The Dappled World*. Cambridge, UK: Cambridge University Press.

Day, T., and H. Kincaid. 1994. "Putting Inference to the Best Explanation in Its Place." *Synthese*, 98, pp. 271–95.

Douglas, H. 2004. "The Irreducible Complexity of Objectivity." *Synthese*, 138, pp. 452–73.

Fine, A. 1986. *The Shaky Game*. Chicago: University of Chicago Press.

Giere, R. 1988. *Explaining Science*. Chicago: University of Chicago Press.

Harman, G. 1977. *The Nature of Morality*. Oxford: Oxford University Press.

Harman, G. 1986. *Change in View*. Cambridge, MA: MIT Press.

Hendry, R. 2003. "Chemistry and the Completeness of Physics," in J. Cachro, S. Hanuszewicz, G. Kurczewski, and A. Rojszczak, eds. *Proceedings of the 11th International Congress of Logic, Methodology and Philosophy of Science*. Dordrecht: Kluwer.

Horwich, P. 1990. *Truth*. Oxford: Blackwell.

Joyce, R. 2001. *The Myth of Morality*. Cambridge: Cambridge University Press.

Kaplan, J. 2000. *The Limits and Lies of Human Genetic Research*. London: Routledge.

Kincaid, H. 2004. "Contextualism, Explanation and the Social Sciences." *Philosophical Explorations*, 18, pp. 201–18.

Kitcher, P. 1978. "Theories, Theorists, and Theoretical Change." *Philosophical Review*, 87, pp. 519–47.

Korsgaard, C. 1996. *The Sources of Normativity*. Cambridge: Cambridge University Press.

Krueger, A. O. 1984. "The Problems of Trade Liberalization," in A. C. Harberger, ed. *World Economic Growth*. San Francisco: International Centre for Economic Growth.

Kuhn, T. 1962. *The Structures of Scientific Revolutions*. Chicago: University of Chicago Press.

Longino, H. 1990. *Science as Social Knowledge*. Princeton, NJ: Princeton University Press.

Longino, H. 2002. *The Fate of Knowledge*. Princeton, NJ: Princeton University Press.

Lucas, R. 1998. "On the Mechanics of Economic Development." *Journal of Monetary Economics*, 22, pp. 3–42.

Mackie, J. L. 1977. *Ethics: Inventing Right and Wrong*. London: Penguin.

Maddy, P. 1997. *Naturalism in Mathematics*. Oxford: Oxford University Press.

McCloskey, D. 1994. *Knowledge and Persuasion in Economics*. Cambridge: Cambridge University Press.

Putnam, H. 2002. *The Collapse of the Fact-Value Dichtomy and Other Essays*. Cambridge, MA: Harvard University Press.

Sklar, L. 2000. *Theory and Truth*. Oxford: Oxford University Press.

Stich, S. 1990. *The Fragmentation of Reason*. Cambridge, MA: MIT Press.

Timmons, M. 1999. *Morality without Foundations*. Oxford: Oxford University Press.

Williams, B. 1985. *Ethics and the Limits of Philosophy*. Cambridge, MA: Harvard University Press.

Williams, M. 1996. *Unnatural Doubts: Epistemological Realism and the Basis of Skepticism*. Princeton, NJ: Princeton University Press.

INDEX

Lightning Source UK Ltd.
Milton Keynes UK
UKOW04n1508300813

216270UK00001B/14/P